AF446823

Property Tax Preferences for Agricultural Land

Property Tax Preferences for Agricultural Land

Edited by
Neal A. Roberts
and
H. James Brown

LINCOLN INSTITUTE OF LAND POLICY

LandMark Studies
ALLANHELD, OSMUN Montclair **UNIVERSE BOOKS** New York

ALLANHELD, OSMUN & CO. PUBLISHERS, INC.

19 Brunswick Road, Montclair, New Jersey 07042

Published in the United States of America in 1980
by Allanheld, Osmun & Co. and by Universe Books
381 Park Avenue South, New York, New York 10016

LIBRARY OF CONGRESS CATALOGING IN PUBLICATION DATA

Main entry under title:

Property tax preferences for agricultural land.

   Papers presented at a symposium in Cambridge,
Mass., June 12–13, 1978.
   Includes bibliographical references and index.
   1.   Real property tax—Addresses, essays,
lectures.   2.   Land use, Rural—Taxation—
Addresses, essays, lectures.   3.   Tax credits—
Addresses, essays, lectures.   I.   Roberts, Neal
Alison.   II.   Brown, Howard James, 1940–
III.   Lincoln Institute of Land Policy.
HJ4165.P76   336.22'5   79-52473
ISBN 0-916672-32-8

Printed in the United States of America

Contents

TABLES AND FIGURES

CONTRIBUTORS

Professor Neal A. Roberts
Osgoode Hall Law School
York University
Toronto, Canada M3J 2R5

Professor H. James Brown
Department of City Planning
Graduate School of Design
Harvard University
Cambridge, MA 02138

John S. Banta
Director of Planning
Adirondack Park Agency
Ray Brook, New York

Professor John Brigham
Department of Political Science
University of Massachusetts
Amherst, MA 01002

Robert E. Coughlin
Vice President
Regional Science Research
  Institute
Philadelphia, PA 19101

Professor David E. Dowall
City Planning Department
University of California
Berkeley, CA 94720

Robert Gloudemans
Senior Research Associate
International Association of
  Assessing Officers
1313 East 60th Street
Chicago, Illinois 60637

Thomas F. Hady
Deputy Director
Community Resources
Economic Development Division
U.S. Department of Agriculture
Washington, DC 20250

Professor Helen F. Ladd
Department of City Planning
Graduate School of Design
Harvard University
Cambridge, MA 02138

ONE

The Big Giveaway Called
Differential Assessment*

NEAL ROBERTS

Introduction

A designer of tax legislation is in many ways an urban planner.[1] Since tax policy, and particularly property tax policy, is rarely neutral in its impact on urban form, the tax lawyer or tax policy advisor should consider the urban impact of possible tax changes.[2] This chapter will explore one such recent change in North America—the creation of preferential agricultural assessments—and will show that its professed land use rationale does not stand up well under analytical scrutiny. It will also suggest that tax policies which appear to have land use justifications are often based on political distribution choices quite divorced from their land use impact.

The last decade has seen three fundamental changes in the nature of the property tax system in North America:[3] the increasing popularity of the split rate, with different sorts of

*A longer version of this chapter has appeared in the *Journal of Urban Law and Policy.*

1

property uses being given different tax rates; the widespread introduction of "circuit breaker" legislation which defers or lowers the tax paid by certain needy or elderly property owners; and the increasing popularity of a system that gives agricultural or open space landowners a tax deferral or outright subsidy at the expense of local or state taxpayers. Each of these developments has been heavily criticized by economists and tax experts, but this criticism has had little or no impact on public decisions.[4]

To analyze these developments we must turn first to the traditional view of the property tax—a tax on wealth which derives its positive attributes from ease of administration and fairness in taxing material assets. In earlier decades it was difficult to determine the market value of properties; thus tax differentiation between similar classes of property was tolerated because of this general imprecision in computation. Whether this disparity was occasioned by kind-hearted assessors who liked old ladies in old houses and disliked businesses or was created by conscious instructions from city councils, with the invention of the computer the anomalies in assessment became much easier to spot. With the invention of computerized mass appraisal systems, they also became easier to rectify. The traditionalist's dream of a real market price determined on a yearly basis became a reality. But just when uniformity was within the grasp of state and provincial legislatures, they started to introduce a new series of legalized inequities between different property owners. The traditionalists have had to face the fact that many people view the property tax not as a wealth tax but as either a benefit tax (related to the amount of local services received) or an income tax (which should be related directly to income derived from property or employment but not from the capital value of the property itself). If one holds this view, the tax can be manipulated easily to benefit certain groups whose income is low despite their ownership of valuable assets or who engage in some type of socially desirable activity with those assets.

This chapter will analyze preferential agricultural assessments—in terms of a rational economic model of the property tax posited on the traditionalist view, and then as measured against the backdrop of the widely held view of the property tax as an income or benefit tax—in relation to the recent plethora of differential assessment programs. However, we first need to consider the possible impact of property taxes on landowners.

The links between tax policy and urban form need careful analysis by public decision-makers. An analysis of any particular tax policy should consider four crucial factors: (1) the impact of the tax policy in terms of equity; (2) the administrative feasibility of the tax policy; (3) its ability to carry out the tax's expressed purpose; and (4) its successful integration with other government policies.

Tax impact on owners of peripheral land

A good deal of irony is present when a farmer who owns land at the edge of a growing city must pay property taxes that are used to service his land, thereby making it more valuable and raising his taxes. While he owns increasingly more valuable land, he may object to a wealth tax which falls most heavily on the one type of asset he owns. His complaints and those of other owners of this type of asset often manage to convince lawmakers that such property owners need help. Before turning to the analysis of these schemes, we should examine how the property tax could affect the market in land.

Land value is derived from the value of the product and not vice versa. Agricultural land derives its value from the value of the products that can be produced on the land; the land purchaser evaluates the present value by determining its discounted future returns. Urban land derives its value from the return that urban uses can bring. Since agricultural productivity is typically much lower than urban productivity, agricultural land is typically less valuable than urban land. If there is an increased demand for urban land, some rural land can be converted to urban use. Between the areas that are agricultural and have no possible potential for urban use and those that are already in urban use, areas exist where investors will be willing to pay a price greater than the current agricultural use value but less than the value for urban uses. This area is located at the periphery of towns and cities.

The purchaser of this land expects the land to have a later urbanized value greater than its capitalized current use value, so he pays a price which includes the current value plus a sum he thinks equivalent to at least the minimum future value of the parcel for urban use. That is, he discounts the future capital value back to a present capitalized sum. His rate of appreciation in this calculation should equal the rate of return for other comparably risky investments. This determination is of course complicated but would include three major variables:

1. The intensity of future developments, based on locational advantages of the land which will save transportation costs between it and other urban parcels.

2. The timing of expected future development, which is also determined by location, since parcels closer to the urban area would be expected to be urbanized sooner, as would parcels close to freeways or major streets.

3. The holding costs between the purchase time and the time of future use.

The primary holding costs for landowners are interest and property taxes. The interest will usually be set at the time of purchase and not vary over the term. The property tax presents the buyer with more of a problem because he must approximate the *total* taxes for the period of time he is going to hold the land and discount that sum back to a present value. He then subtracts that value from his present valuation of the future urban use value of the land. This simply means that fringe land purchasers, all things being equal, pay less for land with a big property tax bill than they would for parcels with a small tax bill.

What is important to note about this investment process is that price decisions based on the property tax liability will have only a minor impact on the particular owner's decision of when to sell. In one jurisdiction the land investor determines what the future development potential is but pays a low price because of a high tax liability, while in another he pays more for the asset because of lower tax liability; in either situation his calculation of when to sell and when he has made sufficient return may come out the same. Since a market exists with many investors, each of whom enters the land market at a different time and with a different set of future expectations, the functioning of the market will be similar in both situations.

The only impact the property tax has on the price or decision to sell is if tax liability changes during the course of the holding period. This of course does not explain why some development occurs at the edge of a built-up area and some scattered out from the edge. The explanation of this phenomenon hinges not on the individual decisions of investors in the area but on the general price levels that result from their trading. The closest land with the highest expectation of immediate development will have the highest price. The land a little more distant may be equally ready for development, if people are willing to pay the extra transportation costs to it, but will have a lower selling price than the closer site. If the price differential between the two sites equals the capitalized transportation costs, a developer will be indifferent between building at one site or another. Thus he will shop within the entire area, and if he finds a site that he thinks is underpriced, he will buy and develop. In such a situation the developers would gain by jumping beyond the immediate fringe and developing on the cheaper land.

We now have some idea about the impact of the property tax and can preview the arguments for and against the preferential treatment of this peripheral land. The argument against a preference is based on the traditional view of the property tax as a wealth tax. A preference destroys the equal treatment of equally valuable real property assets and results in a shift of taxpaying from those who own the preferred asset to those who own other types of real property. Those who use this rational analysis point out that the shift is usually from owners of undeveloped agricultural, commercial and industrial land to homeowners or renters.

The argument in favor of the preference is based on the popular view that the property tax should be related to income capacity, not to the wealth of the individual. Thus a system that taxes at use value is to be preferred, and one that serves other social purposes is even more beneficial. The lower tax payments for the owners of agricultural land is advocated to allow them to continue with their land-rented occupation of farming, to maintain food production on the land, and to provide aesthetically pleasing vistas.

The rational analysis

Differential assessment schemes in the United States and Canada come in a variety of guises that have been explored extensively elsewhere.[5] There are basically three types of schemes. The first, called pure preference, puts a lower assessment on the land than it would command in the marketplace and then applies the normal property tax rate to this value. (This lower value is based not on the capital wealth value of the parcel but rather on its use value. In the case of farmland the value would be the income potential of the land as farmland, not as residential or shopping center land.) The second scheme, called deferred preferential assessment, again taxes the owner on the use-value assessment but keeps track of the difference between the taxes paid at this lower rate and those that would be paid if the land were assessed at its true market value. If the use of the land changes, the scheme provides for recapture of these lost revenues, usually with some type of cutoff on the recapture of perhaps 10 years. The third type of scheme, contract preferential assessment, is the same as the deferred scheme except that the landowner must contract to keep his land in a certain use for a certain time period; the contract provides for penalties as well as recapture if the use of the land is changed. The rationale given for these schemes by their legislative supporters is that they are either providing a needed subsidy to farmers or they preserve agricultural and open space land. We can now analyze this preferential system using the four criteria mentioned previously.

1. EQUITY IMPACT

In the eyes of the law and most economists, property tax is supposed to be a tax on wealth, and all owners of real property wealth are supposed to be treated uniformly. This principle was enshrined in many state constitutions; thus the introduction of differential assessments often required a constitutional amendment. A system of differentiating between real property owners on the basis of the character of their use therefore creates a shift in a tax burden between taxpayers — based not on their wealth, as expressed in the market value of their property, but on the particular type of real property owned. As the property tax is decreased on the fringe landowner, tax revenue must be derived from other property taxpayers within the jurisdictions. Since commercial and industrial taxpayers can, to some extent, pass their property taxes on to the consumers of their products, the burden will be borne by those consumers and the residential renters and owners. Where there are recapture provisions, this shift will, to some extent, be lessened at some future date. In those areas that use state subventions, the shift is from landowners to state income taxpayers.

2. ADMINISTRATIVE FEASIBILITY

Differential assessment schemes are difficult to administer. For one thing, the setting of a current use value divorced from market value is in itself no simple task. The lack of a market standard results in many possibilities for arbitrary valuation, with no easy way for other taxpayers to verify an assessment. The arbitrariness can either be of a local and ad hoc nature or actually codified in the assessment procedures themselves. For instance, is the argicultural use value intended to be the value of the land divorced from its location, or do truck farmers located near urban centers pay more than truck farmers located 200 miles from those centers? Does farm use mean the highest potential use such as mushroom farming or tobacco growing, or does it mean some lesser crops? Strange standards have been set in order to give some uniformity to the process, such as Maryland's rule that farm use value means farm use value as wheat-growing land.

A second aspect to the administration of the schemes is equally troubling: the size of the tax payment that is being shifted from one set of landowners to another set of property owners. Unless the market value and use-value figures are kept up to date, which happens rarely, the magnitude of the preference is kept hidden. Thus the legislators who start the programs have no idea of costs relative to supposed benefits.[6]

Finally, the tax collector or assessor who must police the system is put in the unenviable position of having to make sure the landowners still keep the land in open space or agricultural use. In many Ontario municipalities the towns need a series of employees to follow up zoning changes and building permit issuance and track down parcels which have changed from agricultural to urban use. Municipalities that do not make such an effort often lose the additional tax revenues for a number of years until the assessor finds out about the zoning change.

3.   SUITABILITY

The differential system is supposed to have two purposes: to subsidize farmers and to reduce the conversion of rural land to urban uses. The schemes do provide a one-time benefit to current land investors who purchased their land based on a discounted future tax liability, which they either do not have to pay or which they can pay later under much more favorable terms. This one-time capital increase will be reflected in the current price of the land. For the owner who continues to hold the land, this gain will be shown in his decreased holding costs each year. After the initial windfall, however, new purchasers of such land will not derive any benefits from the program because they will have to pay a purchase price that reflects this lower holding cost.

The second rationale for the program is that it will reduce the conversion of open space land to urban uses, since the converted land is mostly owned by farmers who have a liquidity problem in coming up with yearly tax bills and are forced to sell at inopportune times. While this may be true in some circumstances, it does not resolve the question of whether the purchaser would change the parcel to an urban use or whether the sale would in any way speed up "sprawl." The new owner might continue the agricultural use while waiting until he could put in a more intensive urban use, and he might, as is frequently the case, rent the land back to the farmer from whom it was purchased.

The most important point to make here about the farmer with a cash flow problem is that, in the traditional view, the property tax is not intended to be a neutral tax having an equal impact on the income earned each year from the asset. The tax is on wealth that may include the discounted future development value of the parcel. The owner who has trouble raising the tax money can either sell to another investor or get a loan. Presumably, the local government could simply take mortgages from the landowners in lieu of their tax payment. Moreover, it should be kept in mind here that the landowner is frequently not a farmer but rather an investor.[7]

The whole rationale against conversion seems to be based on the premises that (a) sprawl is an inefficient method of developing cities and (b) the land that is preserved under the differential system will be the farther-removed parcel, rather than the closer-in one at the edge of existing development. While there is some research to indicate that there are costs associated with leapfrog or scattered development, the case is by no means clear.[8] Further, even if the first premise is accepted, most of the schemes do not focus the tax break on this scattered belt of land, but allow it for the entire range of land between urban and completely rural areas.

We now address the question of whether there can be any possible economic land use rationale for the programs, and only three arguments appear to be valid. The first is that the tax subsidy to the landowner will make agricultural use more profitable and encourage him to stay with that use rather than to switch to the urban use which does not get this special subsidy. The trouble with this argument is that urban use is usually many times more profitable on a per-acre basis than agricultural use, as is reflected in land prices that are five to ten times higher for the urban use. Thus the tax subsidy would seem to be a marginal incentive for the landowner.

A second argument is that since the lower taxes will be capitalized by the landowners, they will result in higher land prices which more nearly approximate the final development potential of the land. That is, the prices throughout the investment period will start and stay higher. Since this is so, the argument can be made that the developer who is willing to skip out to a farther-removed site will be less likely to find parcels that have been underpriced. If property taxes are low, any faulty forecasting in the timing of development will not be magnified by the fact that the current low selling price will include the discounted amount of expected low property taxes. This argument does have some merit, but it hardly seems worth the large tax burdens that are being transferred to the consumer and residential landowner.

The third possible rationale has to do with decision-making by the local government involved. The argument can be made that if the actual tax shift is kept visible each year, the local government can evaluate the positive benefit provided by the open space within the municipal boundaries. As the land becomes more valuable, the tax shift will increase, and if the visibility is high enough, the local politicians may be forced to put restricted land into nonagricultural uses. In jurisdictions without such differential assessment, the local governments can downzone or slow development without realizing the full extent of the costs to the community of the open space. That is, the municipality can easily ignore the hypothetical tax revenues that a developed piece of property may bring, but it is more difficult to ignore the lost taxes that must be borne by other taxpayers when the parcel is taxed at a figure below its present market value. Again, there is some merit in this argument, but the current system of hiding the lost taxes or paying the lost taxes out of state income tax funds hardly makes one sanguine about the impact of the program.

4.   INTEGRATION OF TAX AND THE LAND USE SYSTEM

Finally, one must look to the relationship between this tax preference and other government policies. As we have seen, there is little reason to believe that the program would prevent urbanization. If there were such an impact, one would want to focus it on specific land that planners think is "premature" for development. The government would have to develop mechanisms to differentiate between the land which is close in, for which earlier development might be desired, and the farther removed land, which would best be left undeveloped for the present. The preference scheme would have a positive impact, therefore, only if it were localized on the area where skip development might take place. Few jurisdictions even attempt to mesh preferences with land use controls, and virtually all state preference schemes give close-in owners the possibility of taking advantage of the preferences.

From this brief discussion it appears that the agricultural preferences that have gained such widespread popularity are in fact ineffectual tools which would stand little chance of being suggested by rational cost-benefit decision-makers. The fact that the schemes have in fact spread like wildfire must be explained by other rationales.

An historical and political analysis

This fundamental change in the legal structure of the property tax can be understood only with reference to historical factors and political trends with no basis in a rational economic analysis. To conclude simply that the public, and in particular the environmentalists who side with the agricultural groups, have been duped into a giveaway program for which there are no clear apologies does not explain the development of this particular public policy. The change can only be understood in light of the fact that the property tax has never been considered a straight wealth tax by either voter or politician, and that the differential assessment schemes, while representing a fundamental change in the de jure structure of the property tax system, do not in fact represent a de facto change.

The first perspective one can explore is historical. At the turn of the century, for every 1,000 people that were added to a growing city only 10 acres of land were needed. By the time of the Depression, the city needed about 30 acres of land for each 1,000 new residents. Thus the geographic area for potential urban development was relatively limited, and possible tax losses due to assessment at agricultural rather than urban rates were relatively insignificant. With the coming of the post-World War II suburbs, however, land consumption increased tremendously, so that by the 1970s the same new 1,000 residents use more than 200 acres of land. The area of potential urbanization is dramatically larger than it was before the war.[9]

Against this backdrop we can examine the taxation of vacant land. The important point to remember is that the property tax is not one unified tax on wealth but, to quote Professor Netzer, "an incredibly complex collection of taxes with literally thousands of local variations."[10] Most of the tax base is residential and business property; for the nation as a whole, vacant lots and acreage and farms comprise only some 10 percent of the base.[11] Historically, farmland and vacant land have been taxed at a dramatically lower tax rate than have other types of property. In 1962 while the average tax rate for the United States was 1.4 percent of market value, agricultural and farmland was taxed at .9 percent of market value.[12] This was accomplished by setting the assessments on farmland at an appreciably lower figure. Such de facto assessment differences are characteristic of the property tax, where businesses tend to receive higher valuations than residential uses, and farmland receives lower valuations than either.

Commentators have been calling for a reworking of these administratively created differentials for many years, but only recently the technology of assessment has made spectacular changes possible. Since owners of different kinds of property are treated differently, a movement has started toward uniformity, with computerized appraisal systems being introduced to correct the often erratic nature of local assessment. Along with this trend, the political choice underlying the de facto differentials came clearly into the open. Local tax assessors did not discriminate against the businesses and for the open land-owners simply on their own whim. Indeed, assessments reflected the wishes of local governments who had made specific distributional choices. The criticisms of differential assessments, as well as the general movement toward a uniform property tax base, reflect an attempt to impose a rational economic solution in place of the local governments' political conclusions concerning what the impact of the property tax should be in specific communities.

If one follows either the political debate or the discussion of the tax differentials in the popular press, one notes two additional justifications for the schemes.[13] The first concerns transitional equity. Many decision-makers feel that the recent spate of no-growth land use controls has imposed inequitable losses on land investors, which requires some compensation. As we saw earlier, if the land investor invests with a 10-year potential development period and discovers that it is in fact a 20-year period, the present value of the land will decrease. A change to use-value taxation for all present owners represents a compensating increase in capital value. While many would object that this is certainly a most clumsy instrument to accomplish such a compensatory payment, it is at least a scheme that requires a minimum of administration.

The second justification concerns popular myths surrounding land ownership, urban design, and tax policy. It should be clear that the property tax is not neutral and in fact is used to distribute wealth among various segments of the property-owning and -using public. To the extent that land is perceived to be owned by "farmers" and to the extent that farmers are perceived to be worthy of positive wealth transfers, the differentials make political sense even if they do not make economic sense. The myth, however, is that development land is all owned by farmers, and that they would not give up this land unless they were forced to by high property taxes.[14]

The ensuing chapters will attempt to explore these issues from the variety of viewpoints suggested above. In the second chapter. Dr. Helen Ladd will lay out the economic considerations underlying the policy. In Chapter Three the actual land market impact of the policy will be explored both theoretically and empirically by Dr. Robert Coughlin. Turning from the economic to the legal, John Banta, an attorney, will explore some of the legal and administrative problems of implementing the preferential schemes. In Chapter Five, Dr. John Brigham will turn to the political underpinnings of this particular public policy. Finally, the development of the policy and the variety of viewpoints for public policy analysis will be evaluated by Dr. David Dowall.

Notes

1. N. A. Roberts, "The Tax Lawyer as Urban Planner," *British Tax Review* (1975): 345.

2. A recent recognition of the need for such an examination can be found in President Carter's executive order which now requires an assessment of the urban impact of each new federal policy and proposal. One example cited to urge the need for this measure was a tax proposal to abate taxes for construction of industrial plants which, it was claimed, would further hasten the movement of industrial jobs away from urban centers.

3. See A. E. Peterson, ed., *Property Tax Reform* (John C. Lincoln Institute, 1973) and Ronald B. Welsh, "Property Tax Developments: Modernizations and Classification in Site Value Taxation," *National Tax Journal* 29 (1976): 322.

4. Ibid. See primarily J. Shannon, "The Property Tax: Reform or Relief," p. 25, and H. Aaron, "What Do Circuit Breaker Laws Accomplish?" p. 53.

5. See Robert J. Gloudemans, *Use Value Farmland Assessments: Theory, Practice, and Impact* (Chicago: International Association of Assessing Officials, 1971); Thomas F. Hady and Ann G. Sibold, *State Programs for the Differential Assessment of Farm and Open Space Land* Ag. Econ. Report 256 (Washington: U.S.D.A., 1971); Council on Environmental Quality, *Untaxing Open Space* (Washington, 1976) (includes complete bibliography).

6. J. T. Henke, "Preferential Property Tax Treatment of Farmland," 53 *Oregon Law Review* 117 (1974); E. E. Adamson, "Preferential Land Assessment in Virginia," 10 *University of Richmond Law Review* 111 (1970); B. A. Currier, "An Analysis of Differential Taxation as a Method of Preserving Open Space" (unpublished; on file U.S.C. Law Center, 1978).

7. N. A. Roberts and H. J. Brown, *Land Owners at the Urban Fringe* (Discussion Paper D-78-10, Graduate School of Design, Harvard University, 1978).

8. Council on Environmental Quality, *The Costs of Sprawl: Environmental and Economic Costs of Alternative Development Patterns on the Urban Fringe* (Washington, 1971).

9. Charles Sargent. "Land Speculation and Urban Morphology" in *Understanding and Managing Metropolitan Environments in Flux*, (1976).

10. Dick Netzer, *Economics of the Property Tax* (Washington: Brookings Institution, 1966), p. 1.

11. Ibid. p. 18.

12. Adamson, op. cit. *Supra* n. 5 at p. 122, quoting Netzer, ibid.

13. See W. Farr, "The Property Tax as an Instrument for Economic and Social Change," 9 *Urban Lawyer* 447 (1977) at 453: "Whether or not local governments initially intervened to create these tax preferences for homeowners and farmers, changing the system would clearly generate significant political, economic and social repercussions."

14. See Ibid, note 7.

TWO

The Considerations Underlying Preferential Tax Treatment of Open Space and Agricultural Land

HELEN F. LADD

Introduction

Starting with Maryland in 1956 and most recently Vermont in 1978, over 40 states have adopted legislation allowing certain agricultural and open space lands to be assessed at current use value rather than market value for purposes of local property taxation. While state provisions vary from essentially no restrictions on eligible landowners to requiring a long-term contract, the laws all legitimize the use-value approach to assessment of open land.[1] The purpose of this chapter is to evaluate the use-value approach using the three standard criteria of tax analysis: efficiency, equity, and administrative feasibility.[2] In addition, principles of fiscal federalism are used to determine whether the financing of use-value assessment should be a state or local responsibility.

The major purpose of taxation is to shift resources from the private to the public sector; by definition, therefore, taxes impose a burden on private sector taxpayers. In addition, a tax often imposes an extra burden on the economy, referred to as the "excess burden" or the efficiency loss of the tax, reflecting the tax-induced distortions in private sector decision-making. A tax with no distorting effects is said to be efficient or neutral. Provided that private sector markets are otherwise efficient and ignoring equity considerations, a tax structure should be designed to minimize distortion of private sector decisions, thereby minimizing the excess burden of taxation. Although nonneutral taxes can sometimes be justified on efficiency grounds as an offset to nontax inefficiencies in private sector markets, nonneutral taxes are generally considered inferior to direct expenditures for this purpose. These efficiency considerations will be explored in the next section.

Implicit throughout this chapter is the view that equity is a meaningful concept only insofar as it refers to individuals or households. Although it is generally agreed that tax equity requires everyone to pay his fair share, defining fair share is highly subjective. Two generally accepted principles of tax equity have influenced tax policy in the past: the benefit principle and the ability-to-pay principle. According to the former, households should pay taxes in line with the benefits they receive from publicly provided services. The ability-to-pay approach, on the other hand, requires that payments be distributed across taxpayers according to their individual ability to pay and independent of benefits received. In the third section, we will evaluate use-value assessment in relation to these two principles, as well as in relation to a third principle less well accepted but particularly applicable to land value taxation. The section concludes with a discussion of the horizontal equity aspects of optimal tax reform.

A tax is administratively feasible if, at some reasonable cost, taxpayers who are equal according to the tax law are subject to approximately equal tax liabilities. The fourth section compares the administrative feasibility of use-value assessment with market value assessment. Finally, in the last section, we address issues of fiscal federalism, that is, the relationships among governments in a multi-unit system. Three principles of fiscal federalism are used to determine whether revenues lost as a result of use-value assessment should be recovered at the state or the local level.

Economic efficiency: land use decisions

Assuming no other distortions exist in the economy, neutral property taxation of land requires that all land be assessed and taxed on the basis of market value. With this assessment approach, a landowner cannot affect his tax by altering the use of his land: a landowner's property tax burden (ignoring the tax on structures) will be invariant to his current choice of land use.

Many have argued that market value taxation of agricultural or open space land at the urban fringe creates an incentive for landowners who are currently underutilizing their land to shift to a "higher and better," i.e. more lucrative, use. Since the market value of land in urban areas reflects development potential as well as expected income from current use, market value assessment weakens the link between property tax liability and the income from the land that might be used to pay the taxes. Weakening the link, so the argument goes, means that rising property tax liabilities may force landowners to develop their land or to sell it prematurely. Thus, open space advocates who are concerned that market value assessment will have detrimental effects on the preservation of open space justify use-value assessment of agricultural land on the grounds that it reduces forced sales by linking property tax payments to current use income.

Provided that landowners have access to credit, however, the argument that market value taxation of land induces premature sale or development is incorrect. As market values rise, landowners have more collateral with which to borrow, if necessary, to pay the higher taxes. Restricted availability of credit for this purpose is a credit market problem, not a market value assessment problem.[3] Nor does a rising property tax rate generally induce premature development, as is commonly alleged. With full capitalization of property taxes, the current landowner cannot avoid the higher tax burden by selling. While he suffers a capital loss, he cannot avoid that loss by changing his behavior. Again, however, there may be a cash flow or credit problem. If correctly defined as the sum of interest costs (both explicit mortgage payments and implicit opportunity costs) and property tax payments, the costs of carrying the land do not increase with higher property taxes; with full capitalization, the higher tax rate reduces property value sufficiently to insure that the sum of interest costs and tax payments remains constant.[4] The breakdown of these carrying costs between tax payments and opportunity costs will be altered, however, potentially placing the landowner in a tight cash position.

To avoid forced sales caused by cash flow problems or, more generally, by lack of access to credit, property owners can be permitted to defer payment of property taxes with full payback required at market interest rates. In this way, the government (either local or state) would be acting as a lender of last resort. How crucial the problem of forced sales is, and thus how necessary the solution of tax deferral, is not currently known. No state now has a policy of the type advocated here in that current rollback periods and interest rates are insufficient to make the deferred payment a loan rather than a grant.[5] The popularity of existing schemes among farmers thus gives a misleading picture of the efficiency benefits gained from tax deferral.

While market value taxation of land (with provisions for tax deferral) is neutral with respect to land use decisions, tax neutrality may not be desirable. A variety of other factors may lead to inefficiencies in land use that could be counteracted, in part, by preferential tax treatment of agricultural and open space land. First, cities might be "too large" because of excessive reliance on automobiles, encouraged by the fact that the private costs of automobile use exclude the social costs of pollution. Second, open space yields public good or externality benefits that are not accounted for by private decision-makers, thereby encouraging developers to underinvest in open space. Third, the irreversibility of the development process raises concern about the ability of the country to meet future food demands. The implicit argument here is that society discounts future benefits at a lower rate than private developers do, thereby leading to overinvestment in development that yields near term benefits and underinvestment in agriculture that yields longer term benefits.[6] Finally, since local public services tend to yield fewer benefits per dollar of market valuation to owners of agricultural or open space land than to other types of landowners, the demand for agricultural or open space land is less than it would be with a neutral expenditure package, thereby resulting in too little vacant land.[7]

In light of these potential distortions, some form of government action might be desirable to increase agricultural or open space land. We will first consider briefly whether preferential tax treatment of such land can change landowner behavior appropriately. While the total supply of land is essentially fixed and therefore cannot be affected by tax policy, the amount of land devoted to any particular purpose, under normal economic assumptions, will vary in response to the net of tax annual rent for that type of use. By lowering annual tax costs of land devoted to agricultural or open space use, the net of tax annual rent increases relative to what it would be under market value assessment, thereby inducing landowners to devote more land to this activity.

Unrestricted use-value assessment, however, is a blunt policy instrument that benefits all eligible landowners in return for a small supply response at the margin. In addition, by lowering the carrying costs of agricultural land, use-value assessment may encourage speculators to hold onto land in the path of development longer than they otherwise would, thereby forcing developers to leapfrog over those parcels. To the extent that urban sprawl imposes social costs, this is undesirable.[8] Moreover, this assessment approach may reinforce the life-cycle factors affecting farmers' decisions to sell by enabling them to hold onto their property long enough to derive the federal tax advantages of selling at death or retirement.[9] While this result may be desirable on other grounds, it cannot be defended in terms of land use policy because the geographic distribution of farmer-owners bears no necessary relationship to desirable development patterns.

By restricting tax reduction benefits to those landowners who contract not to develop their land for a given period of time, laws such as California's Williamson Act attempt to target funds to the marginal decision-makers. As Schwartz, Hansen, and Foin have demonstrated, the Williamson Act provides sufficiently large benefits to induce profit-maximizing landowners to enroll in the program under a wide range of hypothesized land price patterns.[10] Limited actual enrollments, especially in developing areas, reflect overly optimistic landowner expectations of future land prices. As the authors point out, however, even if landowner expectations could be lowered to induce more enrollment, urban sprawl would not necessarily be reduced. Like other forms of preferential tax treatment, the Williamson Act may provide incentives for leapfrog development.[11]

Is preferential tax treatment the best way to achieve land use goals? Incentives provided by tax reductions have the apparent political advantage of minimizing the visibility of their costs. As many public finance experts have pointed out, however, divergences from the normal, accepted structure of a particular tax should be subject to the same critical evaluation as direct government expenditures.[12] In this case, the difference between the yield when all property is assessed at market value and when certain types are assessed at use value constitutes the cost, or the tax expenditure of the incentive program.[13] Although information on the costs of use-value assessment is limited, one study estimated that if Montgomery County, Maryland, devoted the resources it lost from use-value assessment to outright land purchases, it could have bought one percent of the county's farmland.[14] This suggests that the costs of use-value assessment can be substantial, particularly when few restrictions are imposed on participating landowners.

Moreover, tax expenditures in general and use-value assessment in particular have additional drawbacks as policy tools for achieving behavioral changes. First, tax breaks are often blunt instruments unsuited to targeting specific behavior or groups of people; as noted above, use-value taxation is no exception. Second, tax expenditures may have undesirable side effects. The encouragement of urban sprawl when sprawl is not the goal of use-value assessment policy provides an examp[le]. Furthermore, as we will discuss in the next section, use-val[ue] assessment can have adverse distributional effects.

As a general conclusion, direct expenditure rather than tax policy can better achieve land use objectives. Expenditure policy in this instance might include government purchase of development rights or outright purchase of land, both of which could be targeted to achieve the desired land use patterns. Alternatives to the direct expenditure approach include pricing policy, such as raising the private cost of automobile usage to the social cost, and regulation, specifically zoning and land use controls.

In summary, market value taxation of agricultural and open space land (with provision for tax deferral) is neutral with respect to land use decisions. Although use-value assessment might serve to offset certain nontax inefficiencies associated with urban land markets, it is inferior to direct expenditures as a policy instrument for this purpose.

Equity

Can preferential treatment of agricultural land be justified on the grounds that it produces a fairer tax distribution than does taxation at full market value? Although people may disagree about what is fair, certain basic principles have been used to evaluate the equity implications of tax policy. Besides the benefit and the ability-to-pay principles, a third approach popularized by Henry George in the 19th century asserts that taxing unearned increases in value achieves equity. In the following discussion, these three perspectives on equity are used to evaluate use-value assessment of agricultural and open space land. In the final part of this section, the focus shifts from the equity implications of a particular tax policy to the equity implications of changing from one policy to another.

1. BENEFIT PRINCIPLE

Some analysts view the local property tax as a payment for locally provided public services. They argue that since farmers at the urban fringe derive only limited benefits from the public services provided by urban governments, they should receive a tax break compared to other landowners. Moreover, to the extent that a landowner receives benefits from public services in line with his property's current use value rather than market value, the appropriate form of property taxation of farmland is the use-value approach.

This argument can be criticized, first, because it is difficult to justify the property tax as a benefit tax.[15] A majority of property tax-financed services yield benefits to residents bearing little or no relationship to property tax payments; families with the most children attending public school, for example, do not necessarily live in the most valuable houses. Second, the benefit argument has not traditionally been used as the rationale for property tax exemptions.[16] For example, the most widely used property tax exemption, the homestead exemption, provides tax relief to the group of residents likely to benefit most from locally provided public services, i.e. homeowners. Moreover, proponents of the currently popular circuit breaker form of tax relief for low-income elderly households argue on ability-to-pay rather than on benefit grounds, even though elderly households probably receive fewer direct benefits from certain public services such as education than do other taxpayer groups.[17]

2. HENRY GEORGE

Henry George's concept of tax equity is in direct contrast to the benefit approach in terms of the implications for land taxation at the urban fringe. According to George, tax equity is obtained by taxing away "unearned" increases in value.[18] Since rising land prices in urban areas reflect urban growth and public sector investment rather than the investment activity of the landowners themselves, land value increases are unearned and therefore an appropriate basis for taxation. When this equity argument is combined with the neutrality aspect of land value taxation, the implications for property taxation are clear: all land, including agricultural land in urban areas, should be taxed at market value and preferably at a rate higher than other types of property. Thus, the modern followers of Henry George are opposed to preferential tax treatment because reduction of agricultural taxes vis-a-vis other property taxes moves tax policy entirely in the wrong direction.[19]

3.   ABILITY TO PAY

Although some economists still accept the Georgian view of equity, the majority hold a different view of tax fairness. Instead of distinguishing between earned and unearned income, these economists maintain that a tax policy is equitable if it imposes burdens that bear a meaningful relationship to the taxpayer's total ability to pay. Although opinions differ about how ability to pay should be measured and about the optimal progressivity of the tax structure, most economists agree that high tax burdens on taxpayers with low ability to pay violates this principle of equity and that tax reductions to taxpayers characterized by high ability to pay generally are not justified on equity grounds.

In light of the ability-to-pay approach, what can we conclude about the current tax treatment of farmers and other owners of vacant land? First, we should emphasize that this view of equity (and all other views considered in this chapter) is meaningful only in relation to individuals, not to a parcel of property; we therefore need to compare property tax burdens to total household income or wealth, not to farm income alone.[20] Time series data comparing farm and nonfarm real estate taxes as a percentage of personal income, presented in Table 2.1, indicate that by 1971 the property tax burden on farmers was more than twice that on nonfarmers. It should be noted that this occurs despite lower effective property tax rates (tax payments divided by full market value) for farm property than for other types of property.[21]

Some might argue that higher-than-average property taxes in relation to income are a problem of horizontal inequity; that is, of two families with equal income the one engaged in farming pays higher property taxes, thereby violating the principle that households with equal ability to pay be treated equally. The issue of horizontal inequity, however, is largely irrelevant in this context. If ability to pay is measured by income, only the income tax will be horizontally equitable. Assuming that the property tax is here to stay, nothing is gained by basing arguments for reform on differences in tax burdens within income classes, since these differences are inherent in the tax itself. Issues of horizontal equity will be further discussed below in the context of changes in tax policy.

Nor do average tax burdens in relation to income justify general tax relief for farmers on grounds of vertical equity. In the absence of data for the United States as a whole, Table 2.2 provides some suggestive distributional information for a stratified random sample of 1,213 farmers residing in Georgia in 1972.[22] The distribution of farmers across income classes (column one) is bimodal, i.e., while a substantial portion of farmers are poor, another 16.3 percent have incomes over $18,000. Comparing column one to column two, which shows the distribution of a similar sample of 1,036 Georgian home-owners for the same year, indicates that the proportion of farmers in the highest income category is significantly higher than the proportion of homeowners.

In columns three and four of Table 2.2, income is defined as adjusted gross income plus social security and welfare income; in columns five and six the definition of income is broadened to include appreciation in the value of real estate. The narrower definition of income suggests that farmers with low incomes have excessive tax burdens and generally high tax burdens compared to homeowners. When total income rather than money income is used as the denominator, farmers still make higher property tax payments relative to income than do homeowners, but the lower income classes no longer appear as burdened.

This table raises the important issue of the appropriate definition of income for tax policy purposes. The more com-prehensively defined measure of total income is consistent with the Haig-Simons definition of comprehensive income long advocated by many income tax specialists at the federal level. According to this concept of income, annual income or ability to pay should be defined as annual consumption plus the change in the household's net worth during the year.[23] The major implication of this measure for our purposes is that

Table 2.1   **Farm and Nonfarm Real Estate Taxes
as a Percentage of Personal Income (United States)**

Year	(1) Farm Population (percent)	(2) Nonfarm Population (percent)	(3) Ratio of (1) to (2)
1950	3.31	2.58	1.28
1952	3.31	2.53	1.31
1953	3.83	2.56	1.50
1954	4.22	2.69	1.57
1955	4.67	2.69	1.74
1956	4.81	2.74	1.76
1957	5.11	2.85	1.79
1958	4.81	3.07	1.57
1959	5.53	3.07	1.80
1960	5.74	3.23	1.77
1961	5.73	3.44	1.66
1962	5.81	3.48	1.67
1963	5.97	3.49	1.71
1964	6.22	3.47	1.79
1965	5.79	3.47	1.67
1966	5.92	3.48	1.70
1967	6.62	3.57	1.85
1968	7.04	3.26	2.16
1969	7.07	3.35	2.11
1970	7.48	3.46	2.17
1971	7.72	3.62	2.11
1972	6.94	—	—

Note: Column (1) is reported directly from U.S. Department of Agriculture data. Column (2) was obtained by (a) subtracting farm real estate taxes from total property taxes; (b) multiplying by the (estimated) ratio of total real estate to total property taxes—in 1956, 1961, 1966, and 1971 this ratio was calculated as the ratio of the assessed value of taxable real estate to the assessed value of all taxable property, as reported by the Bureau of the Census, and for other years the ratio was interpolated; and (c) dividing by total U.S. personal income less personal income of the farm population.

Source: U.S. Department of Agriculture, Economic Research Service, *Farm Real Estate Taxes* (Washington, D.C.: U.S. Department of Agriculture, March 1974); U.S. Department of Commerce, Bureau of the Census, *Census of Governments*, vol. 2 (Washington, D.C.: GPO), 1972 and earlier issues; U.S. Department of Commerce, Bureau of Economic Analysis, *Survey of Current Business* (Washington, D.C.: GPO, August 1973), pp. 42–43.

Reproduced from R. J. Gloudemans, *Use-Value Farmland Assessments: Theory, Practice, and Impact* (International Association of Assessing Officers, 1974), Table 12, p. 9.

capital gains need not be realized by sale to be considered income since they add to the household's net worth. The fact that a substantial portion of the wealth of low-income farmers in urban areas is in the form of unrealized increases in land value should be irrelevant to their ability to pay, provided that, these landowners can borrow against their land to pay taxes or from the government in the form of deferred taxation.

Data in Table 2.2 clearly demonstrate that money income underestimates the true income of farmers more than of homeowners: the average figures at the bottom of the columns three through six show that total income is four times money income for farmers, compared to 1.15 times money income for homeowners.[24] It should be noted that total income for farmers will be particularly understated in urban fringe areas where land values are growing the most rapidly.

On ability-to-pay grounds, use-value taxation is an inappropriate way to relieve the farmer's tax burden. Even if relief is restricted to bona fide farmers, it is still not well targeted to those with low ability to pay.[25] Under use-value taxation schemes, the farmers who benefit the most are those owning rapidly appreciating land,[26] i.e., those least in need in terms of ability to pay, correctly defined.

A second, perhaps obvious, point is that tax relief directed at owners of vacant land rather than farmers per se is a grossly inefficient means of helping farmers, and again is likely to have perverse equity implications. Owners of vacant land at the urban fringe include corporations, family-held businesses, syndicates, and individuals, many of whom own land for speculative reasons. A study of the California Williamson Act found that ten landholding corporations owned over one-fifth of the affected land, yielding between $4 to 5 million in tax relief each year.[27] Evidence also suggests that when use-value legislation was initially enacted in Maryland, many wealthy individuals began purchasing farmland, making it difficult to restrict participation to bona fide farmers.[28]

In a recent study of vacant land owners in the Phoenix metropolitan area, Brown and Roberts report that for their particular sample 41 percent of the land is held by individuals, 8 percent by family businesses, 31 percent by syndicates, and 20

Table 2.2 Taxes as a Percentage of Income for Georgia Homeowners and Farms

Adjusted Gross Income Categories	Distribution of Observations		Property Tax as a Percentage of Money Income[a]		Property Tax as a Percentage of Total Income[b]	
	(1) Farmers (percent)	(2) Homeowners (percent)	(3) Farmers (percent)	(4) Homeowners (percent)	(5) Farmers (percent)	(6) Homeowners (percent)
Less than $2,000	16.8	3.8	59.7	8.3	6.5	5.2
$2,000–$4,000	12.2	9.2	18.2	2.1	5.4	1.8
$4,000–$6,000	13.2	12.2	11.0	1.4	4.4	1.2
$6,000–$8,000	10.6	16.3	8.9	1.1	4.3	1.0
$8,000–$10,000	8.3	17.6	6.3	1.1	3.4	1.0
$10,000–$12,000	7.8	14.0	6.9	1.0	3.5	1.0
$12,000–$14,000	6.1	11.2	5.4	1.0	3.1	1.0
$14,000–$16,000	4.8	5.4	6.7	1.4	3.6	1.3
$16,000–$18,000	3.9	3.9	6.0	1.2	3.3	1.0
Over $18,000	16.3	6.6	4.4	1.2	2.7	1.1
Average	—	—	17.3	1.5	4.3	1.3

[a]Money income is adjusted gross income plus social security income and welfare income.

[b]Total income is money income plus appreciation in real property value.

Source: Fred C. White, Bill R. Miller and Charles A. Logan, "Comparison of Property Tax Circuit-Breakers Applied to Farmers and Homeowners," *Land Economics* 52, 3 (August 1976), p. 358.

percent by corporations.[29] The authors also indicate that approximately half of the individual and syndicate owners hold the land for profit and only five percent for farming. In addition, over 80 percent of the land in their sample is held with the motive of selling for profit and only 10 percent for farming. Indeed, even within the geographic region of the sample classified as the farming area, farmers represent only 16 percent of all owners and own only 24 percent of all land.[30] Although speculators play an important positive role in the operation of the land market, their equity claim for tax reduction is not strong. Evidence presented by Brown and Roberts on the income and wealth of the sampled Phoenix landowners shows that many of the owners have high income and substantial wealth.[31]

Unlike use-value assessment that extends tax relief to specific types of landowners regardless of their ability to pay, the circuit breaker approach now widely used to protect the elderly and other low-income families could be extended to farmers.[32] Although many variations are possible, the basic idea is to allow low-income farmers a refundable credit on their state income tax for property taxes that exceed a certain percentage of their income.

Even the circuit breaker scheme has serious flaws, however. First, it is based on the assumption that tax payments reflect tax burdens. To the extent that taxes can be shifted from the taxpaying group to another, tax burdens will differ from tax payments. While there is limited leeway for shifting taxes on land from landowners to consumers or other farm workers, capitalization of future tax streams means that original rather than current landowners bear the burden of property taxes. Relief to current landowners may not be justified to the extent that they avoided the tax burden by paying a lower purchase price for the land. Second, income for circuit breaker purposes should be comprehensively defined to include increases in land values. In practice, this would be difficult to do. Finally, within any income group, the most relief goes to farmers in areas with the highest property tax rates and to those with the largest land wealth.

Circuit breakers of all types raise the question of why relief should be related to property taxes in the first place. Although the scheme does not receive high marks as a form of general income maintenance, at least it would be superior to use-value assessment as a means of targeting aid to the farmers most burdened by the property tax.

4.   TAX CHANGES AND HORIZONTAL EQUITY

Pressure for preferential treatment of agricultural and open space lands comes largely in response to the movement for assessment reform. As shown in Table 2.3, even without legal use-value assessment, farmland tends to be underassessed in relation to market value compared to other property types. Somewhat surprisingly, states with use-value assessment legislation actually assessed acreage at a higher percentage of market value than did other states. Underassessment of open space land means that court-mandated movement to uniform assessment will increase taxes on agricultural land vis-a-vis other property types, resulting in substantial capital losses for landowners.

Table 2.3.   Assessment Sales Ratios by Property Type, 1971

	Assessment Sales Ratio			Comparative Ratios	
	(1)	(2)	(3)	(4)	(5)
	All Property	Single Family Residential	Acreage	Acreage/ All Property (Col. 3 ÷ Col. 1)	Acreage/ Single Family Residential (Col. 3 ÷ Col. 2)
26 States without Use-Value Legislation, 1971	26.9	27.6	18.9	.69	.67
24 States with Use-Value Legislation, 1971	37.9	39.5	28.8	.74	.71

Source: R. J. Gloudemans, *Use Value Farmland Assessments: Theory, Practice, and Impact* (International Association of Assessing Officers, 1974), Table 17 and 18, p. 30. Derived from data reported in U.S. Department of Commerce, Bureau of the Census, *1972 Census of Governments*, vol. 2, pt. 2 (Washington, D.C.: GPO, 1973), pp. 34-39.

Enactment of use-value legislation may thus be viewed less as an attempt to give tax relief to farmers and owners of open space land than as a protection against the movement to uniform assessment. Although use-value tax treatment of agricultural land cannot be justified on ability-to-pay grounds in a situation of uniform assessment, assessment reform itself will create horizontal inequities unless mitigated by use-value legislation. For example, consider an investor who decides to buy land and another investor with an equal income who invests in stocks. Presumably the price paid by the land investor reflects the expected stream of future property tax burdens; the expectation of continued de facto use-value assessment would induce him to pay a higher price than if he expected future taxes to be based on market value. Now consider the effects of a shift to uniform assessment. The land investor is suddenly faced with the higher taxes associated with full market value assessment. If he tries to avoid the tax burden by selling, he will have to accept a capital loss equal to the present discounted value of the stream of additional taxes. The investor in stocks, however, is relatively unaffected.[33] Thus, the change in tax policy, by changing the rules in the middle of the game, imposes a burden on the individual who happened to invest in land.

This line of reasoning has strong implications. To the extent that existing tax policies are fully capitalized into prices or wages so that rates of return are equalized, any attempt to correct distortions or to improve the distribution of the tax burden will cause horizontal inequities. These inequities should then be weighed against the gains from tax reform to determine the desirability of the change. Moreover, from a political perspective horizontal inequities make tax reform extremely difficult to achieve in practice.

In a recent paper on optimal tax *reform* as distinguished from *design*, Martin Feldstein has argued that the losers from tax reform should be compensated, thus relieving inequitable tax burdens as well as presumably increasing political support for reform.[34] Since payment of compensation is not generally acceptable, however, can anything be done to reduce the horizontal inequities? In the context of property taxation of open land, the question becomes one of legislative alternatives to legalizing the de facto preferential tax treatment in the face of pressures for assessment uniformity. One possibility that has so far been unacceptable politically is to ignore the horizontal inequities and allow landowners' effective tax burdens to increase.

A second alternative for states without uniform assessment is to enact laws immediately requiring the implementation of uniform assessment at some specified date, that is, postponing the effective date of tax reform.[35] In this way, the present value of the future increased taxes to current owners will be lower because they begin at a later time. In addition, knowledge of the change in legislation will reduce investment in the activity to be more highly taxed in the future, thereby raising before-tax returns to current owners. Note that this second effect requires current enactment with a future effective date, not merely postponed enactment. The point is to provide advance warning so that investors will not make commitments on the basis of incorrect information. In delaying the effective date of uniform assessment, the horizontal equity argument for preferential tax treatment of agricultural land becomes less compelling.

This line of reasoning also has implications for attempts to remove the tax benefits of use-value assessment in favor of some more suitable policy to encourage open space or to provide tax relief for farmers. To minimize the horizontal inequities associated with the increased tax burden of argicultural or open space land, the effective date of the law ending use-value assessment should be postponed.

Administrative feasibility

The third major criterion for evaluating tax policy is administrative feasibility. Even the most equitable and efficient forms of taxation may be undesirable if poorly administered. A tax is administratively feasible if, at some reasonable cost, taxpayers who are equal according to the tax law are subject to approximately equal tax liabilities.

Market value taxation of agricultural and open space land has an administrative advantage over use-value assessment since actual sales data can be used. Even if a parcel has not been sold recently, assessors can compare it to property for which sales data *are* available. Critics of market value assessment point out the difficulties of equitable market value assessment at the urban fringe. They stress the heterogeneity of the land market, the limited number of buyers and sellers, and the role of speculation in developing areas.[36] Beyond these objective problems are the pressures to underassess agricultural and open space land because of the theory that market value assessment imposes excessive tax burdens or has adverse effects on land use decisions.

Use-value assessment does not solve the administrative problems of market value assessment, however, and in fact creates new ones. Use-value assessment rules out the comparable sales approach, since land prices in developing areas will reflect the expected returns from speculation as well as the returns from current land use.[37] This leaves only the capitalization-of-earnings approach, under which a capitalization rate is applied to net earnings from agricultural use of the land. In addition to the standard problems such as determining the appropriate capitalization rate, this approach is complicated by the use of potential net earnings under prudent management rather than actual net earnings. Use of actual earnings rewards inefficiency and provides incentives for minimizing the tax burden by minimizing agricultural income from the land.

States generally provide local assessors with guidelines, usually based on soil productivity ratings, for determining potential farm income. Although a step in the right direction, soil productivity is not the only determinant of potential agricultural earnings; access to transportation and drainage, for example, also need to be considered. Moreoever, soil productivity ratings can be misleading based on incorrect assumptions of the economically most profitable crop.[38]

In addition to practical difficulties, use-value assessment may introduce additional administrative inequities.[39] In many cases, the assessor must determine a particular land parcel's eligibility on the basis of minimum annual income from the required land use. To the extent that two eligible properties are treated differently by the assessor, horizontal inequities occur.

The administrative inequities associated with use-value assessment are impossible to measure empirically because of lack of data. The conclusion that market value assessment is superior to use-value assessment on administrative grounds is thus strictly judgmental. There is little doubt, however, that use-value assessment of open space and agricultural land will, in general, not ease the administrative burden of assessing land in developing areas.

Fiscal federalism

As explained earlier, use-value taxation of agricultural or open space land can be viewed as tax expenditures. This section addresses the question of what level of government should bear the burden of financing use-value taxation, assuming—somewhat contrary to the conclusion of previous sections—such expenditures are worthwhile in terms of efficiency or equity.

Before examining how the standard principles of fiscal federalism apply in this instance, it should be noted that the choice between state and local financing affects the net benefits received by landowners participating in the use-value assessment program. Defining gross benefits as the difference between the property taxes landowners would pay under use-value assessment and under market value assessment assuming a constant property tax rate, net benefits are therefore gross benefits minus the share borne by participating landowners of additional taxes required to finance the gross benefits.

With local governmental financing, to raise any given amount of property tax revenue, the local property tax rate must be increased to offset decreases in the assessed valuation of agricultural and open space land. This increase in the tax rate indicates the impact of use-value assessment on the tax burden of nonparticipating taxpayers in the jurisdiction. The higher tax rate, in turn, reduces the net benefits relative to gross benefits received by participating landowners in two ways: it increases the taxes they must pay on the nonland component of their property; and second, it increases the taxes they must pay on their now lowered assessed value of land. Based on the experience with use-value assessment in 39 Florida counties,

Table 2.4 shows that in some cases the impact on the tax rate can be substantial.[40]

With full state financing of use-value assessment, in contrast, local property tax rates need not be higher than market value assessment since the state government would reimburse local taxing jurisdictions for revenues lost through use-value assessment.[41] Gross benefits to participating landowners exceed net benefits only by the amount of additional state taxes they pay. Farmers in particular—but probably other owners of vacant land as well—are likely to bear a smaller portion of the burden of additional state sales or income taxes than of additional local property taxes. This follows from the wider geographic coverage of state taxes and the lower-than-average ratio of taxable income and consumption expenditures to taxable real estate of farm owners compared to other taxpayers. Thus, for any given

---

Table 2.4.   Percent Increase in Tax Rate[a]

Percent Increase In Tax Rate	Number of Counties	Nonfarm Population	Farm Population	
			Number	Percent of Total Population
0-2	13	2,506,767	9,906	0.4
2-4	8	657,905	11,754	1.8
4-6	9	424,321	13,846	3.2
6-8	4	46,419	5,896	11.3
8-10	2	10,506	7,571	41.9
10-25	2	14,129	1,281	8.3
25-50	1	9,995	101	1.0
50+	0	0	0	—
Total	39	3,670,042	50,355	—

[a]For 39 of Florida's 67 counties.

Source: John C. Keene, "The Impact of Differential Assessment Programs on the Tax Base," in International Association of Assessing Officers, *Property Tax Incentives for Preservation: Use-Value Assessment and the Preservation of Farmland, Open Space, and Historic Sites*, 1975, p. 51.

difference between use-value and market value assessment, participating landowners receive higher net benefits when the revenue loss from use-value assessment is financed at the state rather than at the local level.

We now turn to the normative question of whether state or local government is the more appropriate level for financing use-value assessment. Three principles of fiscal federalism are relevant here.[42] The first states that programs with redistribution as a major goal should be financed at the highest level possible. Second, equitable revenue sources should be used as much as possible, given other policy objectives. And third, for efficiency reasons financing areas should coincide with the benefit areas of public goods.

To the extent that redistribution of income toward farmers or other owners of vacant land is the major purpose of use-value legislation, the burden of such redistribution should be borne by all taxpayers, not just those living in areas with high proportions of participating landowners. In part, this follows from the conclusion that, all other factors held constant, participating landowners receive greater benefits from state-wide than local financing. In the extreme case of a taxing jurisdiction in the path of urban development comprised solely of participating farmers, local financing of use-value assessment would result in no redistribution at all to farmers as a group. More importantly, if relieving the farmer's tax burden is a worthy goal, it surely yields benefits to society as a whole and should be financed by the largest possible group of taxpayers, in this case taxpayers of the state.[43]

The second principle—that the most equitable revenue source be used—relates to issues of fiscal federalism only insofar as some revenue sources are not available at certain levels of government. For example, in a state not allowing local income taxes, this principle translates into an argument for statewide rather than local financing of expenditure programs, provided that statewide financing is based on the income tax and local financing on the property tax. In the case of use-value assessment legislation, the implications of this principle are not clearcut in all states, depending as they do on state and local revenue sources and incidence of the local property tax.

Financing based on a state income tax is undoubtedly more equitable than financing by the local property tax; the equity comparison between state sales taxes and local property taxes, however, is less clear.

The third principle of fiscal federalism, that financing areas should coincide with benefit areas, derives from efficiency principles. By forcing beneficiaries of publicly provided goods and services to pay for those goods and services, even on a group basis rather than on the basis of individual benefits received, public sector decisions will more closely approximate the efficient outcomes of the market mechanism. Implicit is the view that taxpayers jointly decide on expenditure levels based on whether additional expenditures yield benefits in excess of costs.

The question, then, is one of identifying the beneficiaries of the open space encouraged by use-value assessment. In the doubtful case that future food supplies are being preserved by such policies, the beneficiaries are future generations and the program should be financed at the national level. At the other extreme, the primary beneficiaries may be those who live near an open space that, in the absence of use-value assessment, would have been developed; in this case, local financing would be appropriate. A third possibility is that the benefits are metropolitanwide in that use-value assessment encourages a more rational development pattern. While this position is consistent with one of the apparent goals of use-value legislation, even California's Williamson Act does not appear to meet it very well. All that can be said in this regard is that a program that effectively encourages rational development should be financed at the metropolitan or statewide level, not the local level.

Except when open space benefits are primarily local or when statewide revenue sources are clearly less equitable than those locally available, statewide financing of use-value taxation is thus preferable to local financing. An important qualification to this conclusion, however, is the administrative problem of statewide financing. Since state reimbursements to localities are based on the current tax rate times the difference between market value and use-value assessment, local assessors have a strong incentive to overestimate market values. Counteracting this, in part, is the incentive produced by state aid to education formulas that reward jurisdictions with low market valuation. There is no reason to believe, however, that the two incentives will exactly offset each other. Thus, statewide financing places a significant burden on the state tax departments which must oversee local assessing practices.

Notes

1. For a comprehensive description of current practices as of 1977, see Robert Gloudemans, *Use-Value Assessment: Theory, Practice, Import* (Chicago: International Association of Assessing Officials, 1974). Analyses of use-value assessment can be found in Raleigh Barlowe, James G. Ahl and Gordon Bachman, "Use Value Assessment Legislation in the United States," *Land Economics* (May 1973): 206–212; the proceedings of the 1975 Property Tax Forum, International Association of Assessing Officers (1975); and John C. Keene et al, *Untaxing Open Space*, prepared for the Council on Environmental Quality (Washington, D.C.: GPO, 1976.) These references divide differential assessment laws into three types: preferential assessment laws authorizing assessors to assess eligible land on the basis of current use value; deferred taxation laws requiring owners of eligible land who convert land to noneligible uses to pay back all or some of the tax savings; and restrictive agreements requiring the landowner to sign a detailed contract. All three types are referred to as use-value taxation in the current paper.

2. These are the standard criteria of public finance economists. My own understanding of them is derived largely from the teaching and writing of Richard A. Musgrave. See, in particular, R. A. Musgrave and P. R. Musgrave, *Public Finance in Theory and Practice*, 2nd ed. (New York: McGraw Hill, 1976), especially chaps. 1, 9, and 21.

3. This argument has been made by many writers, including Frederick D. Stocker, "The Impact of Ad Valorem Assessment on the Preservation of Open Space and the Pattern of Urban Growth," *Property Tax Incentives for Preservation: Use-Value Assessment and the Preservation of Farmland, Open Space, and Historic Sites*. International Association of Assessing Officers, 1975; Henry J. Aaron, *Who Pays the Property Tax* (Washington: The Brookings Institution, 1975); and Barlowe, Ahl, and Bachman, op cit.

4. This can be seen in the following simple model. With full capitalization of property taxes and a constant income stream before property taxes (R), the value (V) of a piece of property is $V = (R - tV)/i$ where t is the property tax rate and i the rate of return on alternative investments. This simplifies to $V = R/(i + t)$. Thus, carrying costs are $(i + t)V$. With an increase in the tax rate to t', the capitalization process decreases market value to $V' = R/(i + t')$. Hence, carrying costs after the tax increase $[V' (i + t')]$ equal carrying costs before the tax increase $[V (i + t)]$ since they both equal R. Empirical evidence seems to support both the capitalization assumption and the conclusion that property

taxes have no impact on land use decisions. See E. C. Pasour "The Capitalization of Real Property Taxes Levied on Farm Real Estate,' *American Journal of Agricultural Economics* 57 (1975): 539–48 on capitalization. On the impact of property taxes, see Thomas Plaut, "The Real Property Tax, Differential Assessment, and The Loss of Farmland on the Rural-Urban Fringe," RSRI Discussion Paper Series No. 97 (April 1977) and Michael S. Owen and Wayne R. Thirsk, "Land Taxes and Idle Land: A case Study of Houston," *Land Economics* 50 no. 3 (August 1974): 251–260.

5. According to Gloudemans, op. cit., p. 20, in 1974 twenty-four states applied some sort of tax penalty to lands assessed at use value converting to an unqualifying use. Of these, only nine states charged interest penalties, varying from 5 percent in Illinois to 20 percent in Washington and Hawaii.

6. Note that the observation that prime agricultural land is being converted to housing is not by itself sufficient to imply inefficiency in land use. At any one location within the metropolitan area, the least fertile farmland will be converted to housing first, but in some locations, prime farmland will be converted simply because the best agricultural land is the best for housing as well. See George Peterson, "Tax Policy and Land Conversion at the Urban Fringe." Mimeographed (no date), pp. 7–13.

7. This point is made by Peterson, op. cit., pp. 15–16. Note that this is an efficiency argument for preferential tax treatment of agricultural land. This chapter illustrates that the associated equity argument is not compelling. Working in the opposite direction from the inefficiencies listed in the text is the property tax on improvements which decreases the demand for development.

8. The pattern of development often referred to as urban sprawl is not necessarily undesirable. Indeed, it may be the goal of the land use policy. Sprawl imposes social costs only to the extent that the private costs of new development outside the existing public infrastructure do not include the full social costs of public services. For a conceptual discussion of urban sprawl and its complex causes, see Marion Clawson, "Urban Sprawl and Speculation in Suburban Land," *Land Economics* 35 (1962): 99–111.

9. These federal tax advantages can be substantial. See Peterson (no date), pp. 19–21.

10. The present value of the benefits of the contract as a fraction of the initial land price ranged from 13.7 percent to 40.3 percent. The price pattern consistently not favorable to the contract alternative is the peaked price pattern since the contract alternative would rule out sale at the price peak. See S. I. Schwartz, D. E. Hansen, and T. C. Foin, "Preferential Taxation and the Control of Urban Sprawl: An Analysis of the California Land Conservation Act," *Journal of Environmental Economics and Management* 2, 120–134 (1975). In practice, however, a substantial portion of the actual participants own land in areas where the expected time of development is more than 20 years in the future. This suggests the impact of the act on land use decisions is minimal. See David E. Hansen and S. I. Schwartz, "Landowner Behavior at the Rural Urban Fringe in Response to Preferential Property Taxation," *Land Economics* 51, no. 4 (November 1975): 341–354.

11. Hansen and Schwartz, op. cit., p. 351.

12. Stanley S. Surrey, *Pathways to Tax Reform* (Cambridge, Mass.: Harvard University Press, 1973), especially chap. 1. For an application of this

concept to use-value taxation see John C. Keene, "The Impact of Differential Assessment Programs on the Tax Base," in International Association of Assessing Officers, *Property Tax Incentives for Preservation: Use-Value Assessment and the Preservation of Farmland, Open Space, and Historic Sites* (1975), pp. 40–61 and Keene et al, op. cit.

13. The tax expenditure concept is more complicated than this suggests because of the increase in the local tax rate required to maintain revenues when tax expenditures are locally financed. With local financing, the burden shows up as an increase in the local property tax rate with the magnitude of the increase depending on the percentage reduction in the local tax base caused by use-value assessment. See Table 2.4 for illustrative changes in tax rates.

14. Reported in Gloudemans, op. cit., p. 47.

15. See the discussion in Musgrave and Musgrave, op. cit., chap. 14.

16. A different benefit argument has sometimes been used to justify tax exemption for charitable institutions: charitable institutions provide benefits to residents that would otherwise have to be provided by the government and hence, they deserve tax exemption. See John Quigley and Roger W. Schmenner, "Property Tax Exemptions and Public Policy," *Public Policy* 23 (summer 1975): 259–97.

17. In a recent comprehensive evaluation of property tax relief programs for the elderly, for example, Abt Associates lists five objectives of property tax relief, none of which includes reducing taxes to bring them in line with benefits received. See Abt Associates, Inc., *Property Tax Relief For the Elderly: Final Report*. Prepared for U.S. Department of Housing and Urban Development, Office of Policy Development and Research (July 1976).

18. Henry George, *Program and Poverty* (New York: Doubleday and Company, 1914).

19. This view is found, for example, in Dick Netzer, "Property Tax Reform and Public Policy Reality," in Arthur D. Lynn, Jr., ed., *Property Taxation, Land Use, and Public Policy* (Madison: University of Wisconsin Press, 1976), pp. 223–33.

20. This contrasts to efficiency considerations, where taxes and/or public sector benefits may be appropriately examined in relation to property income rather than total household income.

21. Lower effective rates on farm property reflect lower property tax rates in rural areas, plus de facto and de jure preferential tax assessment of farmland.

22. This table appears in Fred C. White, Bill R. Miller, and Charles A. Logan, "Comparison of Property Tax Circuit-Breakers Applied to Farmers and Homeowners," *Land Economics* 52 no. 3 (August 1976): 355–62.

23. This concept is elaborated in Henry Simons, *Personal Income Taxation* (Chicago: The University of Chicago Press, 1938) and more recently in Musgrave and Musgrave, op. cit., chap. 10.

24. While superior to money income as a measure of ability to pay, total income still excludes many income elements, such as imputed income from owner-occupied houses that should be included in a fully comprehensive income measure.

25. Such restrictions are often not part of current practice. Only Texas goes so far as to restrict eligibility to single proprietorships and to persons who

derive more than one-half their income from agricultural operations. The minimum gross agricultural income requirements, found in many state statutes, often fail to separate the bona fide farmers from other owners because they apply irrespective of acreage. See Gloudemans, op. cit., pp. 44–51 and 22–23.

26. The statement in the text does not necessarily apply to states such as California having restrictive agreements. The evidence suggests that the owners of the most rapidly appreciating land generally do not accept the contract. See Hansen and Schwartz, op. cit.

27. The study by Ralph Nader is referred to in Gloudemans, op. cit., p. 45.

28. Gloudemans, op. cit., p. 47.

29. H. James Brown and Neal A. Roberts, "Land Into Cities: The Land Market on the Urban Fringe." Mimeographed (no date), p. 24. The study was a pilot project for a more thorough investigation of landowners in six other metropolitan areas. The authors warn that some sampling problems are present in this initial project.

30. Brown and Roberts, op. cit., pp. 31–33.

31. Ibid., p. 36. I have chosen not to present the actual figures because of the peculiarities of the Phoenix sample. The authors' ongoing research should provide some much needed additional descriptive information on this topic.

32. Currently, Michigan and Wisconsin apply circuit breakers to farmers. Michigan gives farmers a refundable tax credit for property taxes in excess of 7 percent of household income. Wisconsin includes farms and other homesteads up to 80 acres of land in its general circuit breaker program. In other states, circuit breaker relief to farmers is typically restricted to the taxes on the dwelling unit and up to one acre of surrounding land. See Advisory Commission on Intergovernmental Relations, *Property Tax Circuit Breakers: Current Status and Policy Issues*, Report M-87 (Washington, D.C., 1975), p. 17.

33. The investor in stocks will be affected to the extent that increased taxes on agricultural and open space land reduce property taxes on companies in which he holds stock.

34. See Martin S. Feldstein, "On the Theory of Tax Reform," *Journal of Public Economics* 6 nos. 1 and 2 (July-August 1976): 77–104; and Martin S. Feldstein, "Compensation in Tax Reform," *National Tax Journal* 29 no. 2 (June 1975): 123–130.

35. Martin S. Feldstein, "Compensation in Tax Reform."

36. For a discussion of these difficulties, see Frederick D. Stocker, op. cit.

37. Reference to sales of agricultural land in areas not subject to development pressures might be feasible in some instances, but comparisons across areas are difficult.

38. See Gloudemans, op. cit., p. 16, for a discussion of the corn-tobacco dilemma in Maryland.

39. In addition, administrative costs are increased when use-value assessment requires that property be assessed at both use value and market value.

40. The impact of use-value assessment on the local tax rate has been investigated by, among others, Robert N. Schoeplein and Justine D. Schoeplein, "A Second Look at the Impact of Differential Assessment of Farmland and Consequent Tax Shifting: Comment," *American Journal of Agricultur-*

al Economics 54 (November 1972), pp. 679–82; Keene et al, op. cit.; and Hoy F. Carman and Jim G. Polson, "Tax Shifts Occurring as a Result of Differential Assessment of Farmland: California, 1968–69," *National Tax Journal* (December 1971): 449–57.

41. Currently, only three states contribute to the financing of use-value assessment: California, New York, and Vermont. Only in Vermont are the costs fully borne at the state level.

42. For more general discussions of fiscal federalism, see Wallace A. Oates, *Fiscal Federalism* (New York: Harcourt Brace Jovanovich, 1973); and Musgrave and Musgrave, op. cit., chaps. 29 and 30.

43. On the other hand, one might argue that current property owners in taxing jurisdictions with substantial amounts of vacant land receive fiscal benefits in the form of a larger tax base not offset by higher public sector expenditures and that these fiscal benefits could equitably be taxed away. To the extent that the fiscal benefits have already been capitalized into land prices, however, this equity argument does not hold up. This conclusion, however, does not alter the conclusion based on efficiency considerations that expenditure non-neutrality leads to overinvestment in development vis-a-vis vacant land.

Further reading

Atkinson, Glen W. "The Effectiveness of Differential Assessment of Agricultural and Open Space Land." *American Journal of Economics and Sociology* (April 1977).

Bahl, Roy W. "A Land Speculation Model: The Role of the Property Tax as a Constraint to Urban Sprawl." *Journal of Regional Science* 8, No. 2 (1968).

Ching, C. T. K., and Frick, G. E. "A Second Look at the Impact of Differential Assessment of Farmland and Consequent Tax Shifting." *American Journal of Agricultural Economics* 52 (November 1970).

International Association of Assessing Officers. *Property Tax Incentives for Preservation: Use-Value Assessment and the Preservation of Farmland, Open Space, and Historic Sites.* Proceedings of the 1975 Property Tax Forum.

Ladd, Helen F. "The Role of the Property Tax: A Reassessment" in R. A. Musgrave, ed., *Broad Based Taxes: New Options and Sources.* Baltimore: The Johns Hopkins Press, 1973.

Lockner, Allyn, and Kim, Han J. "Circuit Breakers on Farm-Property-Tax Overload: A Case Study." *National Tax Journal* 26 no. 2 (June 1973).

Netzer, Dick. *Economics of the Property Tax.* Washington, D.C.: Brookings Institution, 1966.

Pasour, E. C. "Real Property Taxes and Farm Real Estate Values: Incidence and Implications." *American Journal of Agricultural Economics* 55 (1973).

Shoup, D. C. "The Optimal Timing of Urban Land Development." *Papers Regional Science Association* 25 (1970).

Stocker, Frederick D. "The Impact of Ad Valorem Assessment on the Preservation of Open Space and the Pattern of Urban Growth." *Property Tax Incentives for Preservation: Use-Value Assessment and the Preservation of Farmland, Open Space, and Historic Sites.* International Association of Assessing Officers, 1975.

THREE

The Economic Impact: Differential Assessment and the Conversion of Land to Urban Uses

ROBERT E. COUGHLIN

Introduction

THE LOSS OF FARMLAND TO URBANIZATION

During the last ten years, there have been an increased awareness and concern about the loss of good agricultural land to urban uses; estimates range from 300,000 to 700,000 acres per year of prime land (SCS classes I and II) permanently lost. The total loss over the next two decades would add up to between 1.8 and 4.9 percent of our stock of 385,000,000 acres of prime farmland, which might not appear excessive. When compared with the number of prime acres which are "in reserve" and could be brought into production with some investment, however, the potential losses may be viewed as serious. To replace the prime land which will be lost to urban uses between the present and the end of the century would require that between 11.3 and 31.4 percent of all potential cropland be brought into production (Coughlin et al.).

The seriousness of the problem becomes more apparent if one bears in mind that the change from agricultural to urban use is, for all intents and purposes, irreversible; that the needs of the nation and the world for food and fibre are increasing as population increases, and that productivity increases in agriculture seem to have tapered off. An analysis with a time horizon of 100 or 200 years might result in much more startling numbers—and would be a much more appropriate time scale for analyzing the depletion of a finite resource which is of fundamental importance to life.

At the regional level, the loss of farmland is often much more severe. In Massachusetts, 13.1 percent of all land in active agriculture in 1951 was lost by 1971. California has been losing 41,400 acres per year or about 3 percent every ten years. At the local level, the shift may be most dramatic, yet even there the percent of land actually built on rarely approaches 100 and development typically is scattered over the countryside. As an example, Figure 3.1 shows a county on the urban rural fringe of Minneapolis-St. Paul, 47 percent of whose land was in cropland, orchards and nurseries in 1975 when urban development accounted for 24 percent.

Scattered development has two important effects. First, urban activities intrude in formerly rural areas and make it difficult for farmers to continue to farm. For every acre built on, between 0.46 and 1.30 additional acres are idled (Coughlin et al.). Second, the appearance of the landscape is transformed automatically from bucolic, with its generally positive aesthetic connotations, to urban sprawl, which has few supporters as an aesthetic phenomenon.

GOVERNMENTAL MEASURES TO PREVENT THE LOSS OF FARMLAND
TO URBANIZATION

Many state legislatures and local governments have acted in response to the perceived problem of the loss of agricultural lands. Their approaches have varied, and so far none can claim complete success. In general, their approaches can be classified

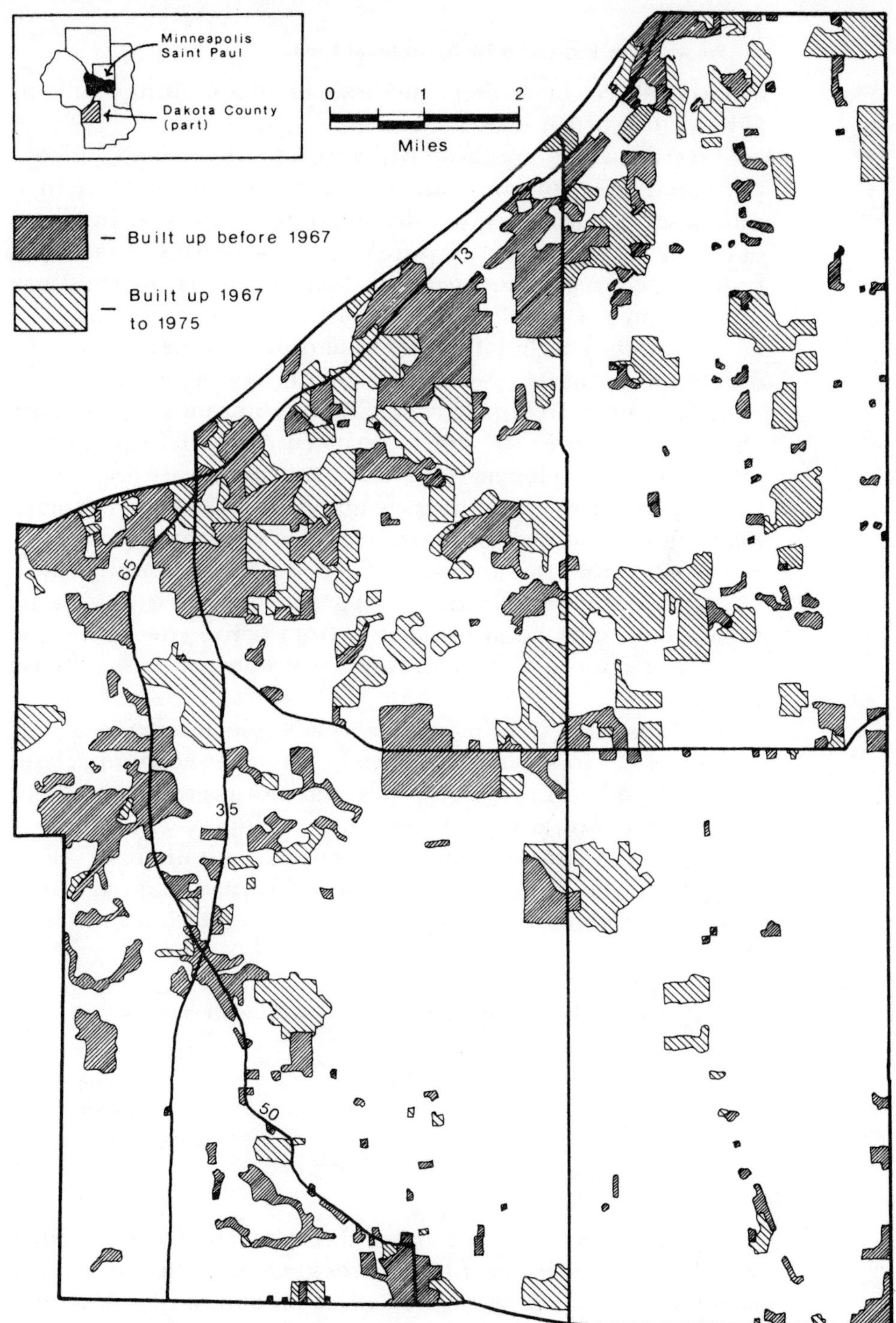

Figure 3.1.   Built-up Area of Dakota County, Minnesota

as either direct or indirect methods for maintaining land in agricultural or rural uses.

Direct measures are those which specify the uses to which a particular piece of land can be put. Direct measures include purchase of fee and less-than-fee interests in land (e.g., purchase of conservation easements, purchase and leaseback or sale with restrictions); regulation (e.g., exclusive farm use or rural use zoning, inverse condemnation, or development permit systems); and the combination of regulation with the exchange of development rights (e.g., transfer of development rights).

Indirect measures provide incentives which make it easier for farmers to continue to farm, or disincentives which make it less attractive for developers to develop. Indirect measures include tax expenditures (such as use-value assessment for property taxation—the subject of this chapter, use-value basis for federal estate tax valuation, and use-value basis for state inheritance tax valuation); tax penalties (such as the Vermont capital gains tax on land sales); and protection from urban effects (such as those protections provided in New York State's agricultural use districts).

It is important to bear in mind that while indirect measures make it easier for a farmer to farm by offsetting economic, land use, and political pressures, they may not prevent the sale of farmland to urban speculators and developers. On the other hand, direct measures, while they help prevent urban development, may do little to make the continuation of farming economically viable. They may save farmland, but not farming.

The Role of Direct and Indirect Measures to Maintain Land in Open Uses

	Make it easier for farmers to continue to farm	Prevent urban development
Direct measures		X
Indirect measures	X	

Thus, by themselves, neither direct nor indirect measures would appear adequate to assure the continuation of a farming economy. Let us now turn to an examination of one widely adopted indirect measure—differential assessment.

The concept of differential assessment

Property taxes on a particular piece of property are based on the property's assessed value, which in turn is derived from its market value. In very rural areas, the market value of a tract of land will be simply the fair market value of the land in its current use, that is, its "current use value." Where there is a possibility that the land could be converted to another use, such as some sort of urban development, the market value of the land may be higher than the current use value. Near rapidly growing urban areas, the market value of land is often several times greater than the current use value. Accordingly, the assessment of such land may rise severalfold.

In addition, the tax rate itself may rise as the urbanizing area needs more revenue to provide water and sewer systems, roads, schools, police and fire protection, and other public services which were unnecessary under rural conditions. As a result, total real estate taxes are typically much higher near urban concentrations than in more remote areas.

As urbanization occurs, taxes rise sharply, but income from rural land uses such as farming and forestry remains relatively constant. Property owners find themselves caught in a "tax squeeze." The tax burden may be so great that the owners of farmland or forest land may no longer be able to realize the minimum necessary economic return. Owners of other land which has low economic productivity, but which is of scenic or ecologic value, also may find that the increased taxes are more than they can afford. As a result, landowners may begin to put their land on the market.

The assertion has been made by others (Aaron, Stocker) that increased taxation need not induce premature sale or development because, as the market value of his land rises, the landowner has more collateral with which to borrow to pay the taxes. This argument appears to be fallacious since borrowing to pay an annually recurring cost will quickly lead to a large cumulative debt which can be paid off only by selling the land itself at a value set by its potential for development.

Differential assessment is designed to reduce the effect of urbanization on assessment of specified types of land, to provide a tax saving for the landowner and, thus, to help make it possible to keep the land in rural use. Under differential assessment, current use value is the basis of assessment.

The difference between use-value assessment and market value assessment—the source of the tax saving to the land-owner in a use-value assessment program—may be very great for a property on the edge of development and insignificant for an outlying property where the likelihood of development is small. This may be seen in Figure 3.2, which gives a generalized view of how this difference and the associated reduction in assessment increases as one approaches an urban center. The figure also indicates that use value varies significantly depending on what crops are suitable to a given piece of land.*

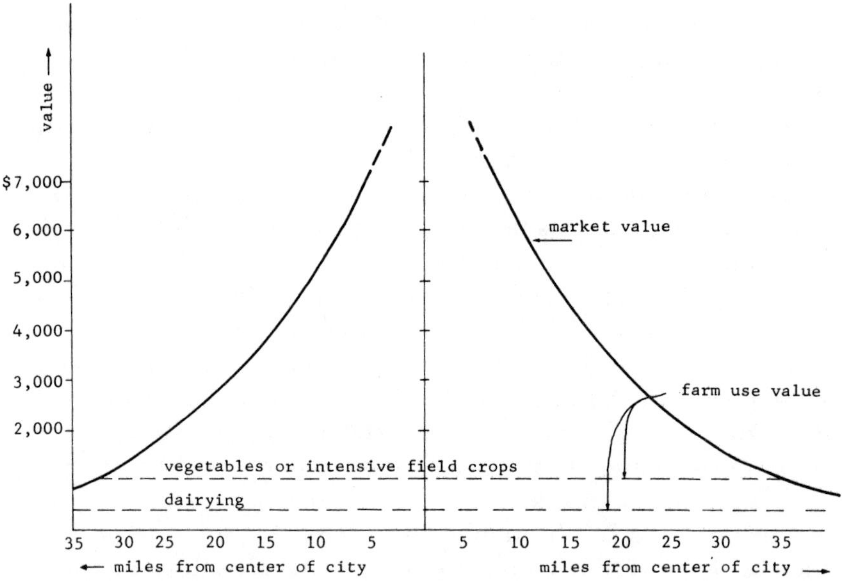

Figure 3.2.   Market Value and Farm Use Value of Land by Distance from Center of a City

*Although classical diagrams indicate a decline in land value for agricultural use with distance from a metropolitan center, in actuality this decline is so slight over 30–50 miles that it is not shown in Figure 3.2.

Tax savings, tax shifts, and participation rates

BENEFITS OF DIFFERENTIAL ASSESSMENT TO PARTICIPATING LAND-
OWNERS

The percent reduction in taxes will not be as large as the percent reduction in assessment. This is primarily because if one landowner is excused from a given amount of taxes, the general tax rate must be increased to the extent necessary to make up that loss in revenue, assuming, of course, that it is necessary for tax revenues to remain constant. There are three factors which determine how great a tax saving will be enjoyed by a landowner under differential assessment:

  i) the difference between the assessed value of the land based on fair market value and its assessed value based on current use value.

  ii) the percentage which the assessed value of all farmland and associated real estate improvements, such as barns, is of the total assessed value tax base before the establishment of differential assessment.

  iii) the percentage which the noneligible improvements or noneligible land in a given property are of the property's total assessed value before differential assessment.

Reduction in assessment. How the probability of development affects the difference between market value and agricultural use value was indicated in Figure 3.2. This generalized diagram is, of course, oversimplified. There is no smooth value gradient rising from the remote hinterland to a major urban center. Instead, the value surface has many minor peaks which are determined by the existence of concentrations of workplaces and residential areas, points of accessibility to the transportation network, and, on a local scale, by frontage on roads, availability of sewers, physical ease of development, and zoning classification. The difference between market value and agricultural value depends on just where in this complex value surface a particular property is located. The total percent reduction for a property owner also depends on what portion of the property is in eligible uses and what portion in noneligible uses. Generally, the reduction in assessment is restricted to land in farming use; improvements and the land on which the farmhouse is located are assessed at market value.

The difference in assessment may not follow the difference in value. Most rural districts have always assessed farm and forest land at use value. Institution of differential assessment laws simply legalized such de facto differential assessment. (See Profesor Brigham's exploration of the historic adoption of differential assessment laws as an attempt to legalize the status quo in the face of the movement toward full value assessment in chapter 5.) Thus, in many cases, little additional reduction in assessment resulted immediately, and in some cases higher assessments were registered. The tax benefits from differential assessment are often in the nature of protection against future rises in assessment due to future rises in the market value of land.

The reduction in assessment also depends on the method used to determine current use value. Few rural properties are sold at current use value since there is almost always some speculative element as well as a value which can properly be ascribed to the probability of development. A capitalization-of-income approach, therefore, is usually employed. In this computation the higher the assumed surplus earned per acre and the lower the capitalization rate, the higher will be the computed current use value of the land. Capitalization rates currently in use vary from 5 percent in Maryland to 10 percent in New Jersey. Therefore, the capitalization computation would indicate that the assessed value of a Maryland farm would be twice that of a corresponding New Jersey farm, even though the two were identical in all other relevant respects.

Tax savings. The reduction in assessments of individual landowners results in a reduction in total assessment within the jurisdiction. To raise the same revenue, therefore, the tax rate must be increased. Thus if all realty in the jurisdiction were in eligible uses there would be no aggregate tax saving to landowners as a class. Percent reductions in assessment, however, would vary depending on the development potential of each parcel. Therefore, participating owners experiencing a greater-than-average reduction in assessment would realize a saving in taxes, and those with less-than-average reduction in assessment would face an increase in taxes. In a taxing jurisdiction where eligible land is a small fraction of the tax base, the tax saving will be shifted to many ineligible landowners. Thus, the reduction in tax payments by participating owners in such a jurisdiction would be nearly proportional to the reduction in assessment.

TAX SHIFTS

Nonparticipants are faced with the same tax rate increase, but for them it is not offset by a decrease in assessment and so their increase in taxes is proportional to the increase in tax rate.

The tax shift is dependent on two factors:

1). The greater the aggregate reduction in assessment for participating properties, the greater the tax shift. The largest reductions can be expected in jurisdictions under intense development pressure; the smallest reductions in jurisdictions with little possibility of development.

2). The larger the percent of the original tax base (assessed at fair market value) which is in participating land, the greater will be the tax shift. That is, if the percent is large, relatively few nonparticipants will have to pay much higher taxes to balance the tax savings of those in the program. Conversely, if a small proportion of the original tax base is assessed differentially, the total tax shift is shared among a great many taxing units and the necessary increase in tax rate is therefore small.

It is difficult to obtain all the data necessary to compute the tax shift in any jurisdiction, but some information is available for California, New Jersey, and Florida.

In California, Gustafson and Wallace report a loss of 4 to 10 percent of county taxes in eight counties, and over 10 percent in three counties (in one of which the loss was 22 percent).

In New Jersey Kolesar and Scholl (1972 and 1975) report that 51.7 percent of all municipalities sampled experienced a tax rate increase of less than 20 percent due to differential assessment, while in Florida, Keene et al. computed that 58.6 percent of all counties (measured by population) experienced a tax increase of less than 10 percent. Thus, on average the shift was rather modest, but a few extreme shifts were noted: in New Jersey, 4.6 percent of all municipalities sampled experienced a tax rate increase of over 40 percent, and in Florida 2.4 percent of all counties experienced a tax rate increase of over 20 percent.

Effectiveness in retaining open space

THE EFFECT OF THE TAX SAVINGS

Although the magnitude of the tax saving may vary widely as was indicated in the previous section, a tax saving generally will be enjoyed by the eligible landowner. This saving alleviates the "tax squeeze" which is experienced by rural landowners, especially on the urban-rural fringe. Under restrictive agreement types of legislation, a change to urban use results in an economic sanction against the participating landowner. But whether the tax savings and the sanctions will actually deter development depends not only on their magnitude but also on the profitability of the land in an eligible use, the personal situation of the landowner, and the demand for developable land. These various factors are diagrammed in Figure 3.3.

The tax squeeze and the decision to sell. For properties in commercial use, such as farming or forestry, the saving in taxes may be enough to increase net income significantly and make it economically feasible to maintain the property in an eligible use. Whether the tax saving makes the property economically viable depends, of course, on the property's basic income-producing capacity: the quality of its soil, the local cost of labor, and many other factors which affect the economics of farming or forestry in the region. For properties maintained as country homes, or "second homes" on large tracts, the tax savings may keep or return the carrying costs to an acceptable level.

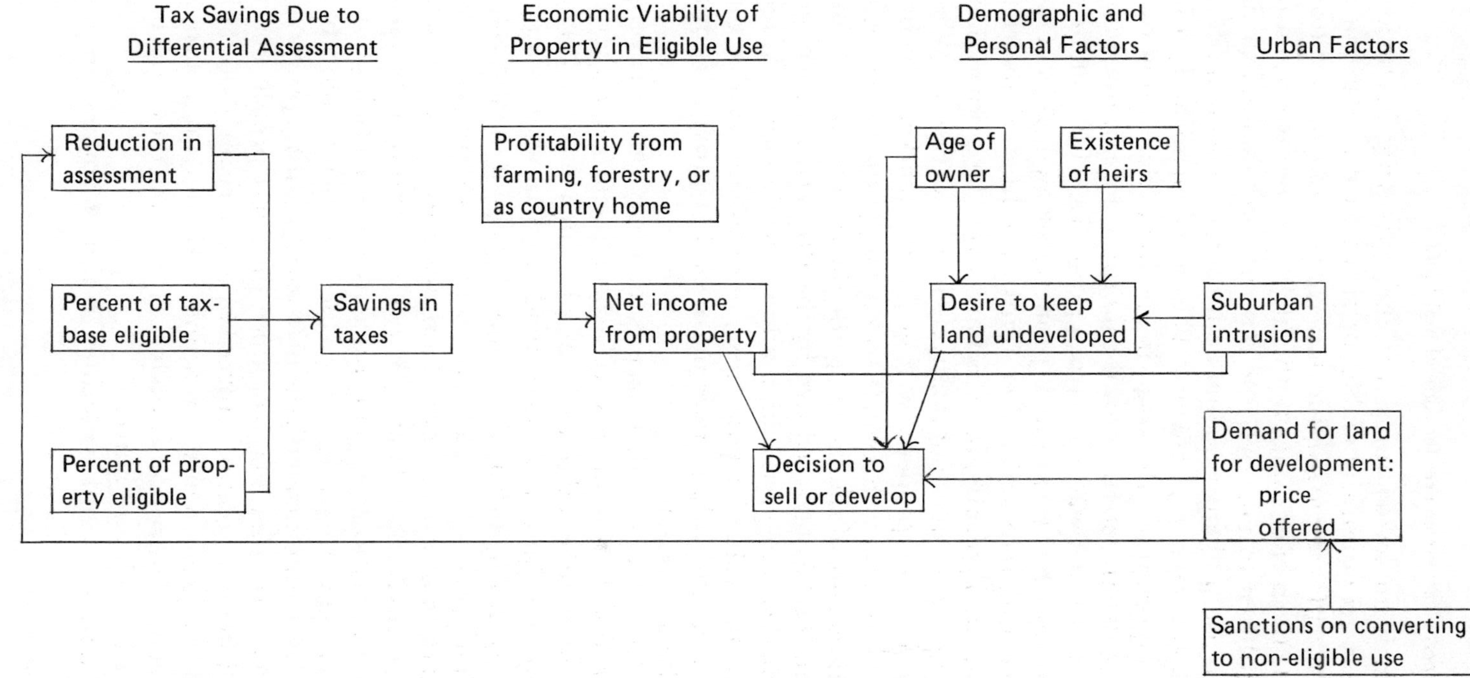

Figure 3.3. Factors Affecting the Effectiveness of Differential Assessment in Retaining Open Space

Even if an owner finds that the tax saving has made his property economically viable, he still may decide to put it on the market for demographic or personal reasons. Such reasons include the owner's stage in the life cycle, the disinclination of any children to take over the property and maintain it in its current use, and any personal desire of the owner to move away or seek a different type of work. The most general, and probably most important, is the stage in the life cycle. Young owners who see deterioration in local conditions for farming may sell and start afresh away from urban pressures, while middle-aged owners may feel tied to their property (Brown, 1978). Owners approaching retirement often think of selling to finance their retirement. Typically, farmers have a very large investment in land and equipment but relatively little cash reserve. When an owner dies, the property will go on the market if he has no heirs; heirs who wish to maintain the property may be forced to sell all or part of the land to settle the estate. The pressure on heirs to sell has been reduced by the 1976 Federal Income Tax Reform Act, however, which allows farmland to be valued at current use value for estate tax purposes, and will be further reduced as individual states make similar provisions in their inheritance tax laws.

Finally, a set of urban factors also may push the property owner to sell. Most of these have been identified as they affect the farmer: increased traffic on roads which makes it dangerous to drive slow farm machinery; inadvertent or willful damage to crops by nearby urbanites; complaints about the use of fertilizers and pesticides; and subsequent local ordinances that forbid farm practices which result in excessive dust, smells, or noise (Conklin and Dymsza, Miner). These problems can make it more costly to carry on rural activities such as farming and forestry, and therefore can reduce the net income from a property. Combined with the increasing political power of new suburbanites, they may convince the rural people that their way of life will soon be replaced by a suburban mode. This so-called "impermanence syndrome" (Alampi) may also affect owners of country houses on large tracts, and these, too, may be induced to sell and move.

Corporations and syndicates owning farmland are less effected by demographic factors or any personal commitment to keep the land in rural uses. For them economic factors are primary. The availability of differential taxation may well encourage the entry of speculators and developers into the rural land market since it reduces the carrying costs of land (though under some market conditions this may be offset somewhat by an increase in land price resulting from capitalization of the tax saving). The change of ownership from farmers and other individuals living on the land to speculators and developers living elsewhere is likely to increase the pace of conversion of land to urban use.

It can be seen that many factors in addition to the property tax burden—or its reduction through differential assessment—can lead to the decision to sell.

Demographic or life-cycle reasons account for the majority of farm sales. Recent studies of New Jersey (Nagle and Derr) and Maryland (Peterson) report that between 55 and 60 percent of sales took place between retirement and death or as part of an estate settlement. Although economic considerations undoubtedly play an important part in many such sales, it is probably safe to say that roughly half of all decisions to sell are made on entirely noneconomic grounds. Whatever effect differential assessment may have will be felt only in some portion of the remaining half of all sales.

Of course, the sale of property does not necessarily lead to a change in use. Rural properties have been sold for generations and remained essentially the same. The critical question is whether there is a demand for developable land and whether the potential developers can offer a higher price than farmers, foresters, or persons wishing to acquire a country estate. In many urban fringe areas, developers can easily outbid those wishing to buy land and keep it in rural use. But where urban pressures are less strong and the quality of the soil is good, the buyer intending rural use may be able to offer a competitive price.

Increase in the price the farmer can afford to pay. It is argued that differential assessment will reduce the gap between what a prospective farmer can bid for land and what a prospective developer can pay. According to well-accepted land economics theory, the tax rate is capitalized into the value of the land: the lower the tax rate, the higher the value of the land. Thus since the effective tax rate is lower under differential assessment, potential buyers intending farm (or other eligible) use can afford to pay more for land and therefore will have a better opportunity to compete with speculators or developers in bidding for land.

The relationship between use value per acre, UV, on the one hand, and net farm income per acre, Y, a capitalization rate, C, and tax rate, T, on the other hand, is as follows:

$$UV = Y \text{ divided by } (C + T)$$

Typically, the capitalization rate lies around the market rate of interest, i.e. 8 percent in recent years, although it may be necessary to adjust it for risk, inflation, and changes in net farm income. For the United States as a whole, in 1971 the average effective real property tax rate was 1.12 percent, although there was considerable variance (Pasour). Assuming T = 2.0 percent, a lowering of the tax rate to 1.0 percent would raise UV by only 1.1 percent. When the federal income tax is taken into consideration, the effective cost of the tax payments—and the corresponding tax saving due to use-value assessment—is reduced, say by twenty-five percent, so that the effect on the capitalized value is further reduced. Hence, one could not expect even a fairly large decrease in taxes to close the gap between use value and market value of land where urban pressures are strong enough to make the development value of land two, three, or many times the use value. Therefore, the farmer looking for land to farm will usually not be able to compete with speculators or developers when bidding for land on the rural-urban fringe. And in fact, under most use-value assessment laws, the speculator or developer can enjoy the same tax saving as the farmer, so long as he keeps the land in agriculture. It has been observed that in New Jersey it is common for owner/speculators to rent out their land to farmers in order to qualify for farm use assessment (Kolesar and Scholl).

Empirical studies of the effect of property taxes. No complete empirical analysis of the effect of property taxes on land sales is known. Plaut (1977) and Berry (1975), however, have analyzed the relative importance of property taxes per acre of farmland on the rate at which land went out of farming in Ohio counties for the periods 1964 to 1969 and 1969 to 1973.

Their studies indicate that the rate of change of land in farms is associated with taxes per acre only in areas of marginal farming productivity (which do not appear to be agriculturally viable in the long run) and in very urban counties (where the effect of taxes is statistically overwhelmed by population increase and the demand for conversion of land). In Ohio counties with highly productive soil, taxes per acre had no discernible effect on the rate of loss of farmland.

Similarly, the loss of land in farms was not found to be significantly related to the effective tax rate (that is, the rate based on fair market value rather than on assessed value). Thus, if the loss of farmland is not related to property taxes, it is unlikely that a reduction in property taxes is likely to have a significant effect.

THE EFFECT OF SANCTIONS

Sanctions such as rollback taxes and interest payments which are imposed in many states provide only a minor deterrent to a change in land use. If no interest is charged, the payment of back taxes cannot be considered a penalty at all from an economic point of view. In effect, such a program simply allows the owner to postpone certain taxes (those in excess of taxes based on the current use value of the land) until his land is developed. Thus, it is equivalent to an interest-free loan to the owner. Imposition of an interest charge itself creates a true penalty only to the extent that the rate is higher than the rate the landowner would have had to pay were he to borrow from a commercial lending institution. The interest rate charged in almost all states is well below recent commercial interest rates.

Of course, an owner may not see the sanctions through the dispassionate eye of an economist. They may add up to a sizeable sum of money and, therefore, could possibly constitute a psychological deterrent to sale for development. Even this possibility is remote under most current state stipulations on years of rollback and interest rate, since tax rates are generally in the range of 1 to 3 percent of market value, the rollback taxes are computed only on the difference between farm value and market value, and interest is computed on the rollback taxes. Assuming the highest tax rate (3%), the highest interest rate (10%), and the highest number of years of rollback (10 years) in existing legislation, the rollback is only about 20 percent of the land value at the time of sale (Table 3.1), but the rollback would be substantially less in more typical situations.

In addition to being ineffective in preventing a change in use, sanctions may discourage some eligible owners from enrolling. Under California's restrictive agreement program, which requires a 10-year contract and imposes substantial sanctions for leaving the program, little land near urban areas or in coastal areas is enrolled (Fellmeth, Gustafson, and Wallace; Hansen and Schwartz; and Keene et al.). Even in New Jersey's deferred taxation program, which has a minimum sanction of a two-year tax rollback with no interest and which involves nearly 95 percent of all farmland in the state, there is some evidence of significantly lower participation rates near major urban areas such as Philadelphia, Trenton, and New York (Keene et al., p. 151).

Although hard empirical evidence is scanty, it appears safe to conclude that as the costs of joining or leaving a program of differential assessment are increased, the participation rate is likely to drop, especially in areas where there is a strong demand for sites for development. Thus, so long as programs are strictly voluntary and are auxilliary to direct land use restrictions, moderately strong sanctions are likely to be accompanied by low participation, especially in areas under urban pressure toward which the land use goals of the program are primarily directed.

Although sanctions on conversion to urban uses may reduce the number of participating landowners and have little effect on sale and development decisions of those who do participate, their inclusion is defensible on grounds of equity. On a change to a noneligible use, they require repayment of at least part of the "tax grant" which other taxpayers have made to the participating landowners to enable them to keep their land in open uses.

Table 3.1 Rollback as Percent of Total Market Value of Land
(Farm Value = 50 percent of Market Value at Year 0;
Appreciation Rate = 10 percent)

Year	Interest at 5 percent		Interest at 10 percent		Interest at 15 percent	
	Taxes 1 percent	Taxes 3 percent	Taxes 1 percent	Taxes 3 percent	Taxes 1 percent	Taxes 3 percent
3	1.7	5.0	1.8	5.3	1.8	5.5
5	2.8	8.5	3.0	9.1	3.5	10.4
7	3.9	11.8	4.3	12.9	4.7	14.0
10	5.7	16.9	6.5	19.4	7.3	21.9

The role of differential assessment

It is clear from the preceding discussion that differential assessment responds only to some of the forces which push land into development, primarily the problem of increased tax burdens due to a rise in land values.

Many other factors may push a rural landowner to sell his land. These include the construction of public and private facilities which disrupt rural areas and induce urban development, the tax burden due to front-foot assessment for new utility lines, the tax burden due to additional urban services within the jurisdiction, damage to crops by suburban neighbors, and local restrictions passed by the votes of new residents on noisy, dusty, or smelly farm practices. A number of indirect, or incentive, measures have been instituted which are designed to deal with such problems. The most all-inclusive is agricultural districting (Conklin and Bryant) which embodies several different incentives. The Oregon land use control system includes most of the provisions of agricultural districting, but sets them in a mandatory system of planning and control.

The New York Agricultural Districting system is a voluntary one. Districts consisting of at least 500 acres in which agricultural uses predominate may be established on the initiative of the landowners with the cooperation and consent of the state. Within such districts local ordinances are generally prohibited which would unreasonably restrict or regulate farm structures or farming practices; front-foot assessments by utility districts are prohibited on agricultural land; local exercise of eminent domain or the intention to advance public funds "for the construction of dwellings, commercial or industrial facilities, [or] water or sewer facilities to serve non-farm structures. . ." must be reviewed by the State Commissioner of Environmental Conservation and the alternatives fully explored. In addition, owners of farmland in a district may apply for differential assessment. But there is nothing in the law to prevent a participating farmland owner from developing his land.

Even though it is economical and practical for a farmer to continue to farm, he would probably succumb to a high offer for his land. Only direct land use controls appear to be sufficient to protect rural uses in the face of such strong market demands. These include regulatory methods such as exclusive agricultural zoning (Coughlin et al.; Lapping et al.; Solberg and Pfister) and the public acquisition of fee or less-than-fee interests in land (Brenneman; Coughlin and Plaut).

Oregon has combined direct and indirect measures. Its Land Use Act of 1973 requires that the state set up planning goals and guidelines, that counties and municipalities prepare comprehensive plans in accordance with them—and that they carry out their plans.

The goals call for dividing all land in the state into:

1) already urbanized land,
2) urbanizable land, i.e., land within a growth boundary which must be drawn around each urban place,
3) agricultural, forest, and open space land subject to the restrictions of the goals, and
4) other rural land suitable for sparse settlement with no or few public services.

Agricultural land (i.e., SCS Class I-IV soils) outside urban boundaries must be zoned for exclusive farm use. This stringent direct control is combined with a number of indirect measures to improve the viability of farming: (1) prohibition of local ordinances which would restrict or regulate accepted farming practices because of noise, dust or odor if such conditions do not interfere with the use of adjacent land; (2) exemptions from assessments and levies of sanitary and water supply districts; (3) valuation at farm use value rather than at market value for state inheritance tax purposes; and (4) qualification for differential assessment. In addition, the Land Conservation and Development Commission (the state agency which administers the program) must review all proposed developments of "statewide significance" (such as transportation and utility systems).

The Oregon system thus places differential assessment in a consistent planning and regulatory framework. Its role is to ensure that property taxes will be as consistent as possible with the income derivable from land in designated and enforceable open uses.

A comprehensive approach, such as that being implemented in Oregon, provides a supportive role for differential assessment, which it appears to be capable of filling. It does not assign differential assessment the major responsibility for maintaining rural uses, a responsibility which differential assessment is incapable of bearing alone.

The role of differential assessment, then, should be a supportive one, acting in concert with a variety of direct and indirect controls set in a consistent planning framework. Some sort of direct land use control, such as exclusive agricultural zoning, should be required whenever differential assessment is applied. This will give assurance that the tax expenditure will be effective in achieving the land use goal and will assure other property owners who bear the shifted tax burden that the participating landowners are not simply receiving a tax break. To further protect the general taxpayer from a tax shift whose land use purpose is not achieved, rollback taxes should be required, and the minimum rollback period should be long (say 10 years), with interest computed at a rate at least equal to the commercial rate (which other taxpayers would have had to pay in order to borrow money to make up the lost revenues).

As part of a comprehensive approach to land use control, differential assessment can be a useful tool, required for fairness to the restricted landowner, and whose cost may be justifiable in terms of the achievement of land use objectives. By itself, however, differential assessment may provide a tax saving to participating landowners but will be ineffective in maintaining agricultural and other open uses in the face of development pressure.

References

Aaron, Henry J. *Who Pays the Property Tax, A New View*. Washington: The Brookings Institution, 1975.

Alampi, Philip (chairman). Report of the Blueprint Commission on the Future of New Jersey Agriculture. Trenton, New Jersey, 1973.

Bailey, Mary. "Farm Real Estate Taxes—1975." RET-16. National Economic Analysis Division, Economic Research Service. Washington, D.C.: USDA, March 1977.

Barlow, Raleigh, and Alter, Theodore R. *Use Value Assessment of Farm and Open Space Lands*. Research Report No. 308. East Lansing: Michigan State University, Agricultural Experiment Station, September 1976.

Barron, James, and Thompson, James. "Impacts of Open Space Taxation in Washington." Washington Agricultural Experiment Station Bulletin 772. Pullman, Wash.: Washington State University, 1973.

Berry, David. "The Effect of Property Taxes on the Loss of Farmland: The Potential Effectiveness of Preferential Assessment." Proceedings of the Meetings of the Middle States Division of the Association of American Geographers, Grand Island, New York. 1975.

Berry, David; Leonardo, Ernest; and Bieri, Kenneth. "The Farmer's Response to Urbanization: A Study of the Middle Atlantic States." Regional Science Research Institute Discussion Paper Series, no. 92. 1976.

Boyce, David E.; Kohlhase, Janet; and Plaut, Thomas. "The Development of a Planning Oriented Method for Estimating the Value of Development Easements on Agricultural Land." Regional Science Research Institute Discussion Paper Series, no. 105. August 1978.

Brown, H. James, and Roberts, Neal A. "Land into Cities: The Land Market on the Urban Fringe." Mimeographed, 1978.

Brenneman, Russell L. *Private Approaches to the Preservation of Open Land*. Waterford, Conn.: Conservation and Research Foundation, 1967.

Bryant, William, and Conklin, Howard. "New Farmland Preservation Programs in New York." *Journal of the American Institute of Planners* 41 (November 1965): 309–396.

Conklin, Howard, and Bryant, William. "Agricultural Districts: A Compromise Approach to Agricultural Preservation." *American Journal of Agricultural Economics* 56 (1974): 607–13.

Conklin, Howard, and Dymsza, Richard. "Maintaining Viable Agriculture in Areas of Urban Expansion." Albany: New York State Office of Planning Services, 1972.

Coughlin, Robert E., et al. *Saving the Garden: The Preservation of Farmland and Other Environmentally Valuable Landscapes Under Pressure of Urban Development.* Report to NSF (RANN). Philadelphia: Regional Science Research Institute, August 1977.

Coughlin, Robert E., and Plaut, Thomas. "The Use of Less-than-Fee Acquisition for the Preservation of Open Space." *Journal of the American Institute of Planners.* November 1968.

Fellmeth, R. *The Politics of Land.* New York: Grossman, 1973.

Gaffney, Mason. "Tax Reforms to Release Land," in Marion Clawson, ed., *Modernizing Urban Land Policy.* Baltimore: Johns Hopkins University Press, 1973.

Gloudemans, Robert. *Use-Value Farmland Assessments: Theory, Practice, Impact.* Chicago: International Association of Assessing Officers, 1974.

Gustafson, Gregory, and Wallace, L. "Differential Assessment as Land Use Policy: The California Case." *Journal of the American Institute of Planners* 41 (November 1975): 379–389.

Hady, Thomas, and Sibold, Ann G. *State Programs for the Differential Assessment of Farm and Open Space Land.* Washington, D.C.: USDA, 1974. Agricultural Economic Report No. 256, Economic Research Service.

Hansen, David, and Schwartz, Seymour. "Prime Land Preservation: The California Land Conservation Act." *Journal of Soil and Water Conservation* 31 (September-October, 1976).

House, Peter. "Farmland and Farmland Owners on the Edge of a Growing City with Special Emphasis on Tax Problems." Ph.D. dissertation, Cornell University, 1967.

Keene, John C. "Differential Assessment and the Preservation of Open Space. *Urban Law Annual,* vol. 14. 1977.

Keene, John C.; Berry, David; Coughlin, R. E.; Farnam, J.; Kelley, E.; Plaut, T.; and Strong, A. L. *Untaxing Open Space.* Washington: Council on Environmental Quality, 1976.

Kolesar, John, and Scholl, Jaye. "Misplaced Hopes, Misspent Millions: A Report on Farmland Assessment in New Jersey." Princeton: Center for Analysis of Public Issues, 1972.

Lapping, Mark B.; Bevins, Robert J.; and Herbers; Paul V. "Differential Assessment and Other Techniques to Preserve Missouri's Farmlands." *Missouri Law Review,* vol. 42 (1977), pp. 369–408.

Locken, Gordon. "Alternative Methods of Estimating the Use-Value of Farmland in New York." Master's thesis, Cornell University, 1976.

Luke, George. ". . . Actively Devoted . . . The First Decade of the New Jersey Farmland Assessment Act." New Jersey Agricultural Experiment Station, Cook College. New Brunswick, N.J.: Rutgers University, 1976.

Miner, Dallas. "Farmland Retention in the Washington Metropolitan Area." Washington: Metropolitan Washington Council of Governments, 1976.

Nagle, George R., and Derr, Donn A. *A Preliminary Analysis of the Data on Participation in the New Jersey Farm Real Estate Market, 1966–70.* New Jersey Agricultural Experiment Station. New Brunswick, N.J.: Rutgers University, 1972.

Nelson, Bryan E. "Differential Assessment of Agricultural Land in Kansas: A Discussion and Proposal." *University of Kansas Law Review*, vol. 25 (1977), pp. 215–245.

Pasour, E. C. "The Capitalization of Real Property Taxes Levied on Farm Real Estate." *American Journal of Agricultural Economics* 57 (1975): 539–548.

Peterson, George E. "Tax Policy and Land Conversion at the Urban Fringe." Land Use Center Working Paper 0875–04. Washington, D.C.: The Urban Institute, 1974.

Plaut, Thomas. "The Real Property Tax, Differential Assessment, and the Loss of Farmland on the Rural-Urban Fringe." Regional Science Research Institute Discussion Paper Series, no. 97. 1977.

Roe, Charles. "Innovative Techniques to Preserve Rural Land Resources," *Environmental Affairs* 5 (1976), pp. 419–446.

Solberg, Erling, D., and Pfister, Ralph R. *Rural Zoning in the U.S.: Analysis of Enabling Legislation.* ERS Misc. Publication 1232. Washington: USDA, July 1972.

Stocker, Frederick D. "The Impact of Ad Valorem Assessment on the Preservation of Open Space and the Pattern of Urban Growth." In *Property Tax Incentives for Preservation: Use Value Assessment and the Preservation of Farmland, Open Space, Historic Sites.* International Association of Assessing Officers (1975), pp. 24–35.

Commentary: Tax Policy and Economic Impact

THOMAS F. HADY

Nearly all the states have now adopted some form of differential assessment legislation. This fact has important implications for those of us who study these laws. We need to stop asking whether these laws are a good thing for states to adopt, since the majority of state legislatures has already answered that question. The important question has become, perhaps, "Now that most states have chosen to adopt some form of differential assessment law, how can they improve it?"

This changes the focus of analysis in a subtle but very important direction. It means that the important issues now are, first, to help the body politic identify just what it was trying to do when it passed a differential assessment law, and second, to help determine the best way to carry out that purpose. Once taxpayers are given a preference, it is extremely difficult to take it away. Hence, states which initially passed a pure preferential assessment law commonly have not repealed it. Instead, when they decided they needed recapture provisions to underline the incentives they were trying to produce, they have added some form of recapture tax to their existing law.

Many have expressed doubts, over the years, about the efficacy of differential assessment laws in carrying out land use objectives. In the face of their widespread adoption, however, it will be more useful if we now concentrate our efforts on modifications of the existing laws, which will relieve the worst problems and improve the ability of these laws to achieve their intended purpose.

One problem area is the variation in assessments which can result when value in agricultural use is not defined carefully. Consider, first, the issue of what earnings are to be included in the capitalization. Clearly, earnings from current farming activities should be included. Equally clearly, if one is finding value in agricultural use one does not include potential urban uses. There appears to be a middle category which often is overlooked, however. In areas far removed from urban pressures, it is difficult to explain values of agricultural land on the basis of capitalized current earnings, except at very low capitalization rates. This suggests that there must be an element of expected higher future earnings in agriculture which is being capitalized into the values of land which will be farmed for the foreseeable future. Market values of this land likely will exceed values found by capitalizing current earnings at interest rates bearing any reasonable relationship to market rates. However, this element of farmland values often is not included in formulas for deriving agricultural value. This fact may help to explain the interest of farmers in areas far removed from urban pressures in having their land under the differential assessment program.

There are other issues, also, involved in estimating earnings. What ability should one assume on the part of the operator? What cropping pattern should one assume? In both cases, the answer is the marginal operator or crop. The farm operator who nearly was outbid for land would set the standard for the quality of management. Similarly, the marginal use for land of a given quality would have to dictate earnings.

One other issue concerning the land market perhaps needs some discussion: the incidence of these tax reductions, and the implications of the "new" theories of incidence of the property tax for differential assessment (Aaron; Stam & Sibold, pp. 13–

16).* Fundamentally, the new view says that the tax, to the extent it is general, falls mainly on owners of capital. The tax we are dealing with is not general, in the sense that it falls only on farm property. It has been adopted so widely that it now makes sense to think of it as a general tax reduction for farm property. To the extent that land can be regarded as in fixed supply, under both incidence theories a cut in property taxes on farmland will result in higher returns from that land, and cause its price to be bid up.

This traditional view needs to be qualified, however. In fact, the supply of farmland often is not really fixed in the rural urban fringe. On the one hand, urban developments are using up that farmland at the extensive margin of urban development. On the other hand, in many areas there is land which is submarginal for any kind of agricultural use at present returns. The rise in prices of farmland will be tempered somewhat, to the extent that "new" farmland can be created from former submarginal land, and some of the tax cut for farmland actually will have its incidence on the owners of that land.

Also, substantial amounts of capital are bound up in much farm real estate in the form of buildings, drainage improvements, terraces, and a host of other improvements. Some states exempt buildings from property eligible for differential assessment but none, to my knowledge, tries to exempt other forms of capital embodied in the real estate (Gloudemans; Hady and Sibold). Hence, preferential assessment of "farmland" actually implies some preferential assessment of farm capital. USDA figures show a steadily declining ratio of farm building values to total farm real estate, but by the mid-1970s farm buildings still accounted for around 17 percent of that value (Stam and Sibold, page 51). No figures seem to be available on other forms of capital now embodied in farm real estate.

Assume, for the moment, that a significant proportion of the value of farm real estate is actually made up of capital (other than buildings). Further, continue the classical assumptions of

---

*Mason Gaffney, in his review of the Aaron book, makes the point that the new view has intellectual antecedents going back as far as E.R.A. Seligman and H. G. Brown (Gaffney, p. 130).

perfect factor mobility and fixed supply of land. That means that part of the tax saving resulting from a differential assessment program will raise the rate of return on capital. But capital is perfectly mobile so it will be attracted to the jurisdiction, until the rate of return on capital in that jurisdiction, equals the rate of return elsewhere.

Furthermore, both the shift of farmland into the industry and the shift of capital allow for a possible increase in supply for farm products and lower prices; some of the incidence of a fairly general differential assessment policy for farmland may be on consumers.

The impact of all this theoretical analysis for practical questions of differential assessment is both good news and bad news for advocates of differential assessment. The good news is that whether one adopts the old or the new view of incidence, much of the incidence of a lower tax on land remains on the landowners. As a result, if one provides a preference in taxation for agricultural land one would expect the incomes of current owners of agricultural land to rise. (Presumably, that increased return from land will be capitalized into values and will not have the same benefit for subsequent purchasers.)

It appears, however, that some unknown part of the value of farmland may actually be embodied capital. To that extent, the new view of incidence suggests that differential assessment will have little effect on income of farmers. The initial impact will be to raise rates of return on capital, but the subsequent effect is to attract more capital into agriculture until rates of return are again equalized. In addition, to a limited degree land may be attracted into farming; these developments will tend to limit the effect of the tax preference on incomes of farmers and incentives to preserve farming.

References

Aaron, Henry J. *Who Pays The Property Tax, A New View*. Washington, D.C.: Brookings Institution, 1975.

Gaffney, Mason. Review of Henry J. Aaron, "Who Pays The Property Tax." *Journal of Economic Literature* 23, no. 1 (March 1978), pp. 129–131.

Gloudemans, Robert J. *Use Value Farmland Assessments*. Chicago: International Association of Assessing Officers, 1974.

Hady, Thomas F., and Sibold, Ann G. *State Programs for the Differential Assessment of Farm and Open Space Land*. Agricultural Economic Report 256. Washington, D.C.: USDA, 1974.

Stam, Jerome M., and Sibold, Ann G. *Agriculture And The Property Tax*. Agricultural Economic Report 392. Washington, D.C.: USDA, 1977.

FOUR

Legal and Administrative Implementation

JOHN BANTA

Introduction

About forty-four states have adopted laws for differential assessment of farmland for property taxes.[1] The mechanical details of these programs differ—some defer taxes;[2] some simply value land for taxation based on agricultural use, ignoring other values;[3] some reduce taxes based on special protective agreements between landowner and taxing authority[4]—but in most cases a program is intended to protect productive farmland from conversion to urban uses.

This protective feature of differential assessment, however, has been widely criticized as a failure if it exists alone. This acknowledged inadequacy poses the question of how differential assessment and other protective techniques like public acquisition of development rights, limits on public investment in sewers and other development services to agricultural land, and land use regulation, work together.

Many states or localities are considering these techniques to complement and strengthen agricultural lands protection. In

71

Oregon, for example, exclusive farm use zoning is now linked directly to preferential assessment to protect agricultural uses in rural areas.[5] California has considered comprehensive new agricultural land protection programs in recent legislative sessions.[6] In New York, a program of agricultural districts has sought some control over new public works to help protect agricultural uses.[7] New Jersey is considering a demonstration program for the acquision of development rights on threatened farmland.[8]

The linkages between assessment and other use protection measures vary. Some are formal, others informal. This chapter will survey implementation problems when these programs operate concurrently by examining briefly the preferential assessment system in four states—Oregon, California, New York, and New Jersey—and at particular problems associated with public acquisition, land use regulation, and public works controls.

In theory,a differential assessment program is unnecessary when another protective technique has attributes of certainty and longevity. Property values and resultant taxes will reflect the restrictions on use. Thus the public acquisition of development rights for farmland should have virtually the same property tax impact as implementation of a differential assessment program. In practice the two may interact in other ways. In any event, the valuation technique used for taxation is very important to the landowner if other protection techniques are uncertain and therefore less likely to be reflected in market-based property valuations.

Local governments cannot legally commit their successors to a specific pattern of agricultural use zoning; they can make only limited commitments to future public works programs.[9] This legal impossibility of certainty and longevity is reflected in market prices. This chapter gives particular attention to the special relationship between regulatory controls and differential assessment.

Four examples of differential assessment

The essential ingredients of a tax preference for agricultural land are an eligibility criterion, a method for establishing assessed value that differentiates between eligible farmland and other land, and an information management system to implement the new procedures. Some programs convert tax savings to tax deferral with provisions for recapture of tax benefits, or a penalty, if land is no longer eligible for use-value assessment.

The legislative formulation of these programs will determine key operational characteristics: (1) eligibility criteria; (2) the nature of local tax records—many keep track only of taxable value, making comparisons of market and taxable value over time impossible; (3) the basis for imputing use value to property—the techniques vary among estimates derived from aggregate farm income figures for a state, from soils characteristics, or from actual income or rent; and (4) the degree of revenue sharing from other sources within a jurisdiction, or from other areas.

NEW YORK

Agricultural land in New York qualifies for differential assessment in two ways.[10] First, a farm of ten acres or more, with a gross average income of $10,000 or more over the previous two years, can enter the program, if, in addition, the farmer agrees to retain the land in agriculture for the following eight years. Second, farms within designated "agricultural districts" are eligible for differential assessment.

Over 5.3 million acres of land are now in agricultural districts.[11] In each case, an annual application for agricultural use valuation is required. Outside the agricultural district, the required eight-year agreement must also be renewed each year. The individual farmer outside a district files this agreement with the county clerk. Within an agricultural district, applications are filed with the local assessor. In either case, the local assessor then uses an average value set by the state Board of Equalization and Assessment, multiplied by the number of eligible acres, and an equalization rate to set the annual assessment for eligible land.[12]

Agricultural use-value factors vary for different counties and types of farm uses. Their calculation is based on a statutory formula, relying on USDA income figures interpreted by a State Farmland Evaluation Advisory Committee. The value factors are imputed to farmland, not calculated from individual user experience. They are important for consistency within taxing jurisdictions, but are sometimes difficult to explain as they apply to a particular parcel of land. For instance, farmers in the lower Hudson Valley have questioned the amount of variation between the per-acre valuation figure for their land and land that is more remote from large cities. The value factor can vary from $1000 to $25 in different areas and use categories. Variation within the same use category may be as much as 100 percent from one area of the state to another.[13]

The program is designed with tax deferral characteristics. If a portion of property within an agricultural district is converted to ineligible uses, a rollback tax recaptures the tax benefit from use-value assessment for the previous five years. Outside agricultural districts, in a similar situation, an owner pays taxes on three times the new market value assessed valuation in the first tax year following a conversion to ineligible use.

The law contains a provision for state revenue sharing when an agricultural district is created by the state, but the state has not initiated districts.[14]

CALIFORNIA

Land in California qualifies for agricultural use-value assessment if it is subject to an enforceable agreement (or "Williamson contract") restricting its use to productive agriculture for a ten-year term. Williamson contracts are agreements between landowners and cities or counties. The program has been in effect since 1965.

Qualifying agricultural uses are loosely defined as the "use of land for the purpose of producing plant and animal products for commercial purposes."[15] Differential assessment is the responsibility of the local assessor. The counties and localities responsible for implementation have considerable interpretive discretion.

In most cases, assessors use a capitalization-of-income approach to determine agricultural use value. Income is based on either comparable rents for similar property or the income that might be imputed to the property under prudent management.

Net income for the property is divided by capitalization rate[16] to arrive at use value and is adjusted by a statutory percentage to arrive at assessed value. A floor for the net income figure to be used in this calculation may bet set by agreement of the parties in the Williamson contract.

As a practical matter, many counties use schedules of rent value per acre for different uses and areas of the county. These become guidelines for the county assessor who values the property for assessment purposes. The tax records are kept only for the current assessed value; comparisons to calculate tax differential depend on the last market value assessment indexed to current value using a factor representing average local market price increases.[17]

If an owner follows the terms of his Williamson contract and gives notice ten (or more) years before a proposed change in use, the tax benefit is phased out. If the agreement is cancelled, a lump sum cancellation fee is charged at 50 percent of the new asssessed value. This is roughly equivalent to a five-year rollback of the tax benefit.

California has a tax-sharing program that provides a different per-acre payment from the state to local taxing jurisdictions for three classes of land in Williamson contracts: urban prime land; nonurban prime land; and nonprime land. The state payment cannot exceed a "tax revenue difference" ceiling set by formula.

NEW JERSEY

Farmland in New Jersey is valued at agricultural use levels for property tax assessment if the parcel is at least five acres in area and, in the judgment of the local assessor, was devoted to active "agricultural or horticultural use" for the two previous tax years. Minimal income requirements must also be met.[18] Over 90 percent of New Jersey's farmland is enrolled in the program.[19]

Property owners must apply each year for agricultural use valuation. Local assessors in some areas then conduct on-site visits to verify eligibility. In other areas, eligibility is based on application information, absent obvious misstatements or complaints. Though assessors must record both use and market value, only the use value for properties is reported as a part of the public record.[20]

A State Farmland Evaluation Advisory Committee assists local assessors in setting agricultural use values for farmland. Using net farm income from USDA, the committee derives county income figures and, by formula, allocates county income according to soil types and use categories. They distinguish woodland, permanent pasture, cropland pastured, and cropland harvested. Roughly speaking, woodland serves as a base of measurement, with pasture values increased by a factor of four times the base value, cropland increased by a factor of ten, and cropland harvested by a factor of twenty.[21]

According to one official of the State Department of Agriculture, the resulting figures are very stable over time, but they can vary significantly from county to county for similar soil and use categories. For example, in Burlington County, class A cropland is now valued at $336/acre, while similar land in a neighboring county is valued at $624/acre.[22]

In New Jersey, a change to urban uses results in a two-year tax rollback for the portion of the land converted to ineligible uses. The rollback is based on the difference between taxation at the new market value and agricultural use values for the property in the year of conversion. Cessation of agricultural uses is also technically subject to tax rollback, but this provision is apparently not strictly enforced.[23]

OREGON

Land qualifies for farm use valuation for the property tax in Oregon if it meets either of two tests. First, land that is zoned exclusively for farming (exclusive farm use or "EFU" zoning) is automatically valued at farm use if the EFU zone is consistent with a county plan and the land is used for farming. Other uses permitted within the EFU zone—schools, churches, forestry, farm dwellings and so on—are specifically excluded from the definition of farm use, and from the preferential tax treatment.

Second, land in farm use for the preceding two years for which the owner can demonstrate a certain minimum gross income for three of the preceding five years is also eligible for farm use valuation for assessment. Once qualified for special assessment under this section, the assessment will continue until the land no longer meets the qualification criteria.[24]

No application is required for differential assessment in an EFU zone; only the initial application and reports of potentially disqualifying actions are required in non—EFU areas. In either case, the program is administered by the county assessor.[25]

Valuation criteria are based on broad statutory guidelines that effectively force capitalization of farm income as the principal valuation technique. Techniques are somewhat similar to California—comparable rental data are sought, and if not available, owner/operator net income is calculated for different types of land and uses in the county. The capitalization rate is set by law and certified annually by the State Board of Revenue.

In Oregon, a conversion to an ineligible use results in a ten-year rollback of the tax benefit from differential assessment for land in the unzoned category, along with interest calculated at a rate of 6 percent from the date at which taxes would have been due. If the owner fails to notify the assessor of the ineligibility, an additional penalty of 20 percent of the principal otherwise due is added. Because the tax benefit for these calculations is derived from the new tax, no dual record system is needed.[26]

Debate over these and other differential assessment programs has been dominated by economists.[27] Characteristics like deferral and increased taxes for other taxpayers have been analyzed with a view to efficiency.[28] But now virtually all states with significant agricultural interests have a differential assessment program of some sort.[29] And the variety of techniques employed suggest that no one approach or academic model has been particularly influential. The time for the "how to do it" discussion is largely past.

The effectiveness of differential assessment has also been carefully examined in recent years.[30] In 1974, the federal Council on Environmental Quality reported: "Another land use control which has become popular in recent years is preferential tax assessments for certain types of property . . .

preferential assessment appeals to a wide range of groups . . . however, there is some question as to the effectiveness of preferential taxation in accomplishing [their] desired goals." By 1976, the muted enthusiasm was gone: "A recent CEQ study indicates that, while these laws do lower the cost of keeping the land in open space, they provide insufficient incentives, by themselves, to affect materially the pattern or rate of conversion to other uses."[31]

The important questions about these programs are not how different characteristics like tax deferral and valuation differ from state to state, but rather how they fit with other agricultural lands/management techniques in a state or locality. During the same 1974–76 time period, the number of differential assessment programs in place at the state level increased from 33 to 42 by CEQ's count.[32] They are a factor in rural land planning. With their emphasis on financial incentives, they play an important role in how other management techniques are evaluated.

There is increasing interest in extending other types of land-use management techniques conceived and first applied in an urban context to rural lands. Public land acquisition, easement or development rights acquisition, zoning and subdivision regulation and control of public services are all being tried in different states, including the four states whose differential assessment provisions were outlined.

The particulars of these proposals vary significantly from place to place because of differences in state laws, institutional traditions and legal constraints that have shaped the urban versions of these land-use controls over the last 75 years.[33] In Virginia, agricultural zoning may be limited by court interpretations that limit the maximum zoning "lot" to a few acres.[34] In Florida, state assistance with large subdivisions is encumbered with bureaucratically complex planning procedures. And agriculture is exempt from most types of state land-use regulation according to the state constitution.[35] Other states have their own program idiosyncrasies.

The diversity in this second generation of agricultural land protection efforts has discouraged a national policy response. Only recently has federal policy begun to address the complex questions of interaction between environmental legislation (NEPA, the Clean Water Act, and others), new state and local land-use initiatives, and other programs like the differential assessment laws. But differential assessment is a common thread.

From this perspective the components of a differential assessment program—eligibility criteria, local records and procedures, valuation, and revenue sharing and recapture—are most important for their relationships to these other initiatives. At the practical level, there has always been recognition that zoning maps and building permits may be a useful source of eligibility information. And where preferential tax eligibility is tied to an agricultural district program, as in New York, the resulting district may be a useful contributor to and constituent of a rural planning process as well as the vehicle for the tax benefit.

There has been less attention to other interrelationships between newly emerging rural lands management techniques and differential assessment.

Differential assessment and techniques to protect agricultural uses

Most states that consider differential assessment a strategy to protect farmland treat it as a financial reward for maintaining land in agriculture. A few others treat the benefit as a reward, but with substantial strings attached. California's Williamson contract offers the best-known example. In a very few situations, notably Oregon and some counties in California, differential assessment is considered a consequence of other restrictions on changes from agricultural use.

In states actively seeking to move beyond the simple "reward" approach to agricultural use protection, there are three techniques of particular current interest: development rights acquisition; regulation that prevents changes in land uses without permission from local or state planning officials; and voluntary agreements that depend in part on the tax benefit, or other government benefits, as an inducement to continue certain agricultural practices.

DEVELOPMENT RIGHTS ACQUISITION

There has been intense state and local interest in development rights acquisition to protect farmland over the past year. Pilot projects have been initiated in Maryland, Massachusetts, Connecticut and New Jersey. One program in Suffolk County, New York, is fully underway.

The problem linkage between development rights acquisition and differential assessment for agricultural use is the acquisition transaction which sets an easily measured value factor to be deducted from market value to set agricultural use value. Problems arise because these figures often vary significantly from imputed use values based on capitalization of income. Where valuation is ultimately a decision for the local assessor, as in New Jersey, such a transaction is likely to be influential. Few statutory capitalization-of-income formulas would prohibit using these transactions for use-value guidelines.[36]

The result can be use valuation either higher or lower than the former imputed capitalization of income. The higher the compensation paid for the development rights, the more likely it is that the sales transaction will establish a use value lower than imputed income value. With rights acquisition funds limited, however, there is pressure to use appraisal or purchase techniques that result in the lowest costs. And with regulation a likely alternative to rights acquisition in some areas, there is an incentive to reach accommodations. These lower acquisition prices are likely to result in higher use values for farmland.

In New Jersey, the agricultural use assessment program is enormously popular. The agricultural use valuation program now uses county income figures to impute acreage values to different classes of land. In Burlington County, where three separate demonstration rights acquisition areas totaling over 13,000 acres are located, assessed values for class A cropland (the only land sought in the acquisition program) were $336/acre in 1977. Appraisers for the rights acquisition program valued the properties at an average of $813/acre. The appraisers used standard market comparisons and parcel-by-parcel income calculations to reach their figures. The methodology is shared with all other state agencies such as parks and recreation departments that acquire limited interests in land.[37]

The values for rights acquisition also ran counter to rental charges for farmland in New Jersey. Apparently rental prices for agricultural use are depressed by the desire of land holders to have active farmers qualify the land for differential assessment. Of the three sample acquisition sites in Burlington County, virtually all of the land in two of the sites was held by speculators (two-thirds of the total acreage).[38]

The realization that many farmland owners in the county faced potential 100 percent increases in assessed valuation if the rights acquisition program went forward was a significant stumbling block as the state Department of Agriculture prepared to send out initial bids in February of 1978.[39] The governor's pocket veto of a program extension has mooted the question until New Jersey's next legislative session.[40]

Corrective measures could fix this problem. Changing the appraisal technique used for valuing rights to be acquired would eliminate the objections of farmland owners, if the tax valuations were used but at a 50 to 100 percent increase in acquisition costs. Alternatively, an amendment to the differential assessment statute could make state guidelines for assessment mandatory. But the legislative climate for revisions to the property tax in New Jersey is so hostile that neither the rights acquisition people nor the farm interests want to be the first to suggest this approach.[41]

EXCLUSIVE FARM USE ZONING

Some restrictions on use to discourage urbanization of farmland appear to be necessary ingredients in any farmland protection program, at least at the urban fringe. If development rights acquisition is a popular land-use management technique because it avoids regulation, it is still unlikely to become widespread because of cost. Where it has been tried, the money seems to run out before all neighbors are able to participate.[42] On the other hand, regulation costs the public purse nothing or relatively little where state or local political support makes a program possible.

One state with a program of agricultural zoning is Oregon. The program is remarkable, in part, for the elements of the "urban" zoning model it eliminates or significantly changes. "Exclusive farm use" (EFU) zones are created with heavy emphasis on the problems of a farmer in an urbanizing area. Rather than emphasize a complicated development review procedure, the program attempts to protect agricultural privileges within the zone.

Incentives and privileges to the property owner within an exclusive farm use (EFU) zone. The legislative "Agricultural Land Use Policy" (ORS 215.243) states that since EFU zoning substantially limits alternative uses of agricultural land, incentives and privileges are justified in order to hold such land in EFU zones. In summary, these incentives and privileges are:

Assurance that only compatible nonfarm uses will be allowed within the EFU zone (215.213).

Assurance that all divisions of land resulting in parcels of land less than 10 acres in size will be reviewed and not approved unless found to be in conformity with the legislative intent of the *Agricultural Land Use Policy* set forth in ORS 215.243.

Prohibition against restrictive local ordinances which would unreasonably restrict or regulate farm structures or accepted farming practices because of noise, dust, odor, or other materials carried in the air if such conditions do not extend beyond the boundaries of the EFU zone. This section does not restrict any governmental unit from lawfully exercising its power to protect the public's health, safety and welfare (215.253).

Automatic review by assessor of land to determine if qualified for special farm use assessment. No application required (308.370[1]; 308.397).

No minimum income must be earned in three out of the five preceding calendar years in order to qualify for special farm use assessment (308.372).

No requirement that farmland be used exclusively for farm use in the two years immediately preceding qualification for special farm use assessment (303.370[2]).

No tax penalty when land qualified for special farm use assessment is removed from the EFU zone following an action by the governing body that was requested or initiated by the owner of the land (308.397[2]; 308.399[3][b]).

Exemption for land qualified for special farm use assessment within an EFU zone, from certain special district assessments (sewer, water, solid waste, vector control). The exemption does not apply to farm dwellings and up to one acre around them (308.401).

Farm use valuation for inheritance tax purposes for land qualified for special farm use assessment (118.155).*

Oregon adopted statewide planning goals and guidelines effective January 1, 1975, which state:

Agricultural lands shall be preserved and maintained for farm use, consistent with existing and future needs for agricultural products, forest and openspace. These lands shall be inventoried and preserved by adopting exclusive farm use zones. . . . [43]

Within the nine counties of the Willamette Valley, the state's Land Conservation and Development Commission is monitoring the implementation of this goal through the creation of urban growth boundaries and a requirement that all lands outside the urban growth boundaries that fall into soil Classes I–IV be in an EFU zone.[44] The requirement is flexible, and property owners with lands with Class I–IV soils can rebut the presumption that they should be in the EFU zone by demonstrating that the land is not suitable for farm use and that nonfarm uses will not interfere with adjacent farm uses.[45]

The state implements this goal by monitoring statutory provisions that require cities and counties to change plans and zoning to conform to the goal. The Land Conservation and Development Commission can set compliance schedules, and if a locality fails to meet certain deadlines, the state can sometimes intervene in use change decisions.

In rural counties that were unzoned, Oregon's differential assessment program had high participation levels before the initiation of the new goal for agricultural lands. According to state Conservation and Development Commission staff, coun-

---

*Source: Oregon Department of Land Conservation and Development (3-24-78). For more detailed information of this general explanation, contact the Oregon Department of Revenue.

ties frequently turn first to this data to determine what land should be placed in the EFU zone.

The role of preferential assessment may be illustrated by the terms of a referendum in a county where the pro- and anti-EFU forces were particularly divided. In Polk County, local citizens put the issue to a vote. The referendum was reduced to the question of whether all preferentially assessed land should be in an EFU zone. It failed by a few hundred votes.[46] As a result, the state has instated a compliance schedule requiring an inventory and rezoning of agricultural lands by January 1979— a period for rethinking and reevaluation.

Oregon's differential assessment program for EFU-zoned land is automatic, in part because court decisions require assessors to take restrictive zoning into account when valuing land for tax assessment.[47] Combined with other decisions that shift to the applicant the burden of demonstrating need for a zoning change, these court decisions have had the effect of substantially changing the expectations of a landowner with restrictively zoned land in Oregon.[48] The Oregon model illustrates one important role for differential assessment: to assure that restrictive decisions by public land-use managers —that are largely beyond the control of the landowner—are reflected by the taxing jurisdiction.

Exclusive-farm-use zoning and development rights acquisition represent the no-compensation and full-compensation alternatives to agricultural use protection. Partial compensation through special agreement between landowner and local, state or federal government agencies is an intermediate protection technique that can supplement the fiscal incentive provided by differential assessment.

VOLUNTARY AGREEMENTS

The most interesting voluntary agreements currently relating to agricultural land use are proposed as a part of a $600 million authorization for a Rural Clean Waters program to be initiated by the U.S. Department of Agriculture (USDA).[49] The agreements are intended to improve water quality by reducing erosion and toxic chemical contamination that result from poor agricultural practices. USDA will use cost-sharing agreements initiated through local Soil Conservation Districts to implement pest management techniques. States and regions will monitor these agreements through "Section 208" planning programs instituted under the federal Clean Water Act.[50]

The new program follows in a federal tradition of voluntarism and cost-sharing for rural and agricultural lands management. USDA already protects wetlands through contract payments to landowners.[51] Recent legislative proposals have suggested an agricultural lands study commission to study prototype efforts for an expanded protection program.[52]

Tax-related voluntary agreements for land-use protection of the Williamson type, however, have not caught on. The fiscal incentive has apparently been insufficient to draw landowners into long-term protection agreements. California's program has been most popular where development pressure is lowest; New York's is little used, with farmers preferring the vastly more complicated agricultural district procedure, even in rural areas where it was never intended to be used by its original sponsors.

Another reason for the failure to rush to voluntary tax-related agreements is that there are few differences between the Williamson agreement and a strictly applied differential tax program with no agreement. The authors of *Untaxing Open Space* (Keene et al.) argue that there is no difference attributable to the eight-year contract required in the New York program because the "contract" is never enforced.

There are two differences. First, the locality or the landowner that entered the agreement may go to court to seek specific performance, i.e., to prevent a change in use, or to ensure the tax benefit.[53] But as a practical matter, it is easier for a public body to prevent a change in use through its zoning and subdivision approvals than through court proceedings. Similarly, it is easier for the landowner to deal directly with the assessor in seeking answers to specific valuation problems.

A second and perhaps more important consequence of entering a protection agreement is the awareness of future obligations to maintain the land in agricultural use. Expectations of public approvals and public investments are key factors in the development of rural lands, and the mutual commitment of a public body and private owner may be valuable as a communication device.

A Williamson contract, for instance, must be preceded by the creation of an agricultural preserve by the local government involved; and within two years of the creation of the preserve, the land within the preserve must be zoned for agriculture. Because of strict linkages between local plans and zoning in California, the process for entering a Williamson contract is virtually equivalent to a request for plan amendment and zoning change to an agricultural land classification.[54]

The contract also specifies the permitted uses of the property for its duration. The two parties to the agreement are the local government and the landowner. (It is unclear what rights other landowners in the same preserve have under the contract.) The contracts are automatically extended for an additional year on each annual renewal date unless a notice of nonrenewal is given.

The links between the contract and underlying zoning and land-use plans add to its importance. These links are the tools that a locality uses for easy enforcement of the contract.[55]

In the past, some localities have tried to "zone" formally land by individual agreements. In most cases, the courts have rejected these attempts on a variety of theories. The typical situation involved the payment of a fee to a city for a desired zoning use classification. The Williamson contract illustrates a different twist to the same general principle. The locality, through a tax expenditure, pays a fee for acquiescence in a desired zoning classification for a particular parcel of land.

There are few formal prototypes for partial compensation associated with zoning. Whether a tax expenditure of the sort connected with the Williamson contract could legally be considered partial compensation for the imposition of re-strictive zoning is uncertain. Whether it could be lumped with other compensation like the Clean Waters agreements is even less certain. But the legal question is never reached in the context of a voluntary agreement.

The same question approached from a different viewpoint has led courts in some states to declare invalid any valuation for tax assessment that fails to take into account existing zoning or other land-use restrictions.[56] And payment of taxes based on development expectations is a factor in determining vested rights in some jurisdictions.[57]

New York's agricultural districts differ from California's and provide a voluntary association without any specific agreement regarding use. Nor are they linked directly to local zoning. A somewhat involved procedure of public notice and hearing leads to a plan for an agricultural district, review by county planning authorities, creation of a local agricultural review committee, and filing with the state Agricultural Resources Commission. Each proposed district must contain at least 500 acres.

Measures to protect agricultural use, in addition to differential assessment, include prohibition of local government restrictions on farm structures or practices unless reasonably related to health, and controls on state agency activities that might interfere with agricultural uses within the district. Lands within a district may also be exempted from certain tax levies related to improvements for urban services.

The agricultural districts law has not solved problems of farmland use conversion in areas of intense development pressure like Long Island, where Suffolk County has a multimillion dollar development rights acquisition program underway. The state legislature is also considering a small demonstration program to acquire farmland development rights in other areas.

Suffolk County, New York, offers a sidelight to the differential assessment issue. Prior to selling development rights, farmers in the Suffolk County acquisition program had shown little interest in New York's agricultural district program and attendant tax benefits. After selling development rights, they have sought district status, not for the tax benefit, but to protect themselves from harassing local regulations.[58] Although the county has initiated the development rights acquisition program, other local jurisdictions within the county share aspects of land-use control authority like zoning and nuisance control regulations. The Agricultural District is a relatively simple method to assert agricultural interests and reduce potential conflicts between local and county objectives.

The future of agricultural land management, particularly in

areas under urban pressure, seems to lie in this packaging of voluntary and regulatory programs, with financial and other practical advantages for agricultural landholders to balance inconveniences and inequities. Differential assessment is a key financial factor to weigh in these considerations.

Another aspect of rural lands management that needs further investigation is the role differential assessment (de facto or de jure) plays in increasing or diminishing local support for a broad range of protective efforts for agricultural land uses. If differential assessment insulates the farming community from the costs of sprawl—rising taxes for services—it may artificially separate the farmer from his local neighbors when local land-use management decisions are made. Because individual land-owners and their preferences have enormous influence on regulatory approaches to land-use management, this artificial division of interests would be especially influential.

The evidence to date—widespread popularity, failure as a protective device, relationship to full-value taxation, sensitivity to rights acquisition valuation procedures—indicates that differential assessment is a powerful communication device, even if it is ineffective as a land management technique. The partial compensation role that it might play in variations of the exclusive farm use zoning concept needs further elaboration. Its role in generating high expectations for compensation in the same situation will also be a factor in the rural lands management debate as national policy seeks middle ground between the costs of programs like the Rural Clean Waters effort and the enormous variation in local situations when mixed regulation/compensation programs are considered.

Notes

1. R. E. Coughlin, D. Berry, and T. Plaut. "Differential Assessment of Real Property As An Incentive to Open Space Preservation And Farmland Retention," *National Urban Recreation Study*, vol. 1 (Washington, D.C.: U.S. Department of the Interior, 1978), p. 191.

2. Alaska, Connecticut, Hawaii, Illinois, Kentucky, Louisiana, Maine, Maryland, Massachusetts, Minnesota, Montana, Nebraska, Nevada, New Hampshire, New Jersey, New York, North Carolina, Ohio, Oregon, Pennsylvania, Rhode Island, South Carolina, Tennessee, Texas, Utah, Virginia, Washington. Ibid.

3. Arizona, Arkansas, Colorado, Florida, Idaho, Indiana, Missouri, New Mexico, North Dakota, Oklahoma, South Dakota, Wyoming. Ibid.

4. California, Florida, Michigan, New Hampshire, Vermont. Ibid.

5. Oregon Revised Statutes, Section 308.370 (1977).

6. A major push by Assemblyman Charles Warren in the 1976 legislative session, AB 15, finally yielded to coastal legislation as a legislative priority. Subsequent efforts have been less ambitious in scope, concentrating on issues like inheritance taxes as well as agricultural use.

7. N.Y. Agriculture and Markets Law, Article 25-AA (1976).

8. As of November 1978, the governor's pocket veto of the extension of the demonstration program had postponed further consideration of the various proposals under discussion in New Jersey.

9. In some areas, zoning agreements with a landowner are binding on a successor administration (e.g., preannexation in Illinois); but there are virtually no precedents for unilateral action by a city council preventing future administration from changing zoning to other more lucrative uses. Some jurisdictions do require complicated procedures before zoning is changed, such as amendment of a General Plan (e.g., California).

10. N.Y. Agriculture and Markets Law, Article 25AA, Section 300 et seq. (1976).

11. Personal communication, Henry Stebbins, June 1978.

12. Elaine Moss, ed., *Land Use Controls in New York State.* (New York: Dial Press, 1975), pp. 335–341.

13. John C. Keene, David Berry, R. E. Coughlin, J. Farnam, E. Kelly, T. Plaut and A. L. Strong, *Untaxing Open Space.* (Washington D.C.: Council on Environmental Quality, 1975), p. 337.

14. Personal communication, Henry Stebbins, 9 June 1978.

15. California Government Code, Section 51201. (1976)

16. The capitalization rate has three components: one derived from the yield rate for long-term U.S. government securities and set by the state Board of Equalization; one representing risk, based on location, and land use and characteristics; and one reflecting property taxes. John C. Keene et al. op cit., p. 275.

17. Ibid., p. 279.

18. $500 for the first five acres and 50¢ an acre for each additional acre. "N.J. Farm Life Remains a Lot of Peaches, but Less Cream." *New York Times,* 11 June 1978, p. 8.

19. Personal communication, Thomas Hall, New Jersey Department of Agriculture, 5 June 1978.

20. John C. Keene et al., op cit., p. 148.

21. Personal communication, Thomas Hall, 5 June 1978.

22. Ibid.

23. John C. Keene et al., op cit., p. 158.

24. Oregon Revised Statutes, Sections 308, 370(1), 308, 397 (1977).

25. "Land Use Planning and Tax Assessment." Report of Land Use Committee, p. 38.

26. John C. Keene et al., op cit., p. 206.

27. Or at least an economic perspective. The Keene study, the most comprehensive available, was built on a solid inter-disciplinary team. Other discussions reflect a nonsubstantial economic orientation, for instance, "Property Tax Incentives for Preservation and Use-Value Assessment and the

Preservation of Farmland, Open Space and Historic Sites," the Proceedings of the 1975 Property Tax Forum. This may reflect, in part, the substantial practical problems in applying an equitable use-value standard. This problem is a persistent one for appraisers. I suspect my conclusions regarding compensation and regulation can only complicate further their problems with pressures for further manipulation of the system.

28. These arguments are concisely summarized in Chapter 2 of this volume by Helen Ladd.

29. Only Kansas holds out.

30. Annual Report of the U.S. Council on Environmental Quality for the year 1974 (Washington, D.C.: G.P.O. 1974). Annual Report of the U.S. Council on Environmental Quality for the year 1976 (Washington, D.C.: G.P.O. 1976), p. 96.

32. Ibid.

33. See Richard Babcock, *The Zoning Game* (Madison, Wisc.: University of Wisconsin Press, 1958).

34. D. Godschalk, D. Brower, L. McBennett, and B. Vestal. *Constitutional Issues of Growth Management.* (Chicago: ASPO, 1977).

35. E.g. *Florida Statutes Annotated*, Chapter 380 (1978).

36. See John C. Keene et al. op cit.

37. Personal communication, Thomas Hall, 5 June 1978.

38. Ibid.

39. Ibid.

40. See "New Jersey Farm Life Remains a Lot of Peaches, but Less Cream." *New York Times,* op cit.

41. The property tax question is only a small part of the formidable cost and procedural problems in implementing the development rights program. Ibid.

42. The New Jersey situation is only one example. "Experimental" small-scale programs are being considered in Maryland, Connecticut, Massachusetts, New York, and elsewhere.

43. Oregon Land Conservation and Development Commission. "Statewide Planning Goals and Guidelines." Goal No. 3.

44. Department of Land Conservation and Development. "Common Questions About Statewide Goal #3, Agricultural Lands." (Unpublished, 1977)

45. Ibid.

46. David A. Beltz, "Planning Polk County—A Case History." (Unpublished paper, 1978.)

47. *Martindale v. Department of Revenue*, 5 OTR Ado. Sh. 148 (1972).

48. *Fasano v. Board of Commissioners of Washington County,* 264 Or. 574 (1973).

49. P.L. 95–217 (1977).

50. P.L. 92–500 as amended by P.L. 95–217 (Clean Water Act of 1977).

51. Conservation Foundation *Letter,* "Effort to Save Wetlands Are Bogged Down," (October 1978).

52. H.R. 4569, before the 95th Congress proposed such studies.

53. Governmental agencies are familiar with this type of litigation relating to procurement contracts for goods and services. The principal

complication in pursuing a judicial remedy is the involvement of the courts and court procedures. This might be an advantage when a successor government wishes to avoid political accountability for protective agreements established in earlier administrations. See note 9.

54. California

55. But in the case of Suffolk County, New York, the county does not control zoning in many areas. The "agricultural agreement" may serve a larger function of coordinating zoning adminstered by one jurisdiction and agricultural use policies of another in such a situation.

56. See *Martindale v. Department of Revenue,* op cit.

57. See Reilly, ed., *The Use of Land* (Crowell, 1973), p.

58. Personal communication, John V. N. Klein, 14 June 1978.

Commentary: Legal and Administrative Implementation

ROBERT J. GLOUDEMANS

John Banta addresses the problems of implementing use-value farmland assessment programs, and offers some general conclusions for meeting these problems. In this review I would like, first, to add some specific suggestions for the design of a "good" use-value assessment program and, second, comment on the three extensions or supplements to use-value assessment discussed in the latter half of the paper.

Mr. Banta raises four important problems in the design and administration of a use-value farmland assessment program and discusses how these programs have been faced in several states. The problems are (1) determining eligibility criteria, (2) estimating use-value, (3) deciding whether market values should also be estimated and recorded, and (4) deciding whether tax losses should be reimbursed by the state (or province).

Eligibility criteria differ widely among states and provinces. In any case, legislation requiring only a past history of farm use, a minimum number of acres in "farm use," or a minimum gross income regardless of land area must be regarded as inadequate for ensuring bona fide farm use. One viable alternative is to require an average minimum income or output per acre preferably depending on the productivity classification of the land. Another alternative is to require the owner (or tenant) to derive an average minimum percentage of personal income from the land. These requirements may best be stated on an either/or basis. In setting the requirements it should be remembered that the typical farmer derives almost half of his income from nonfarm sources. It may be sufficient to require the owner to affirm that the necessary criteria are being met, but to submit proof if the land appears underutilized or the assessor is otherwise in doubt.

Estimation of use values constitutes the largest of the four problems from an administrative viewpoint. The problems of use valuation are well known and have been discussed elsewhere in this book. For the most part these problems revolve around the necessity of disregarding sales data which reflect nonfarm influences. This compromises application of the sales comparison approach to value, as well as the income capitalization approach, because of the difficulty in supporting capitalization rates with market data. In addition, lack of sales data, coupled with abandonment of the market value standard in valuation, confounds the monitoring of assessment equity through assessment-sales ratio studies.

In view of these considerations, it seems that the key to consistent, supportable use-value assessments must le in a good set of soil productivity maps. These maps would delineate soil classes or types throughout the state or province in sufficient detail for valuation purposes. Once the soil ratings had been established, regularly updating use values to reflect farm input and output price changes would be a relatively easy task. Soil productivity maps would tend to produce a rational, defensible pattern of use values both within and between counties. Many states have made much progress in the development of such maps.

There is little question that the landowner should be subject to a penalty for converting land enrolled in a use-value farmland program to a nonfarm use. This does not necessarily mean, however, that the assessor should place both a use-value and market value each year on all properties in the program, with withdrawal penalties based on such differences. Dual assessments create much additional work and expense which, in retrospect, are rarely justified in that the "drop-out" rate tends rarely to exceed one or two percent per year. The state of Oregon has discontinued its dual assessment requirement for this very reason. Alternatives to the traditional "rollback" tax are a penalty based on either the market value of the land (as in California and New Hampshire) or the difference between farm and market values (as in Maryland, Oregon and New York) at the time of transition. The penalty can, of course, be graduated to reflect the number of years for which a tax advantage was obtained.

Currently three states (Alaska, California and New York) reimburse local governments and school districts for revenues foregone due to use-value assessments. Only Alaska provides complete reimbursement, "subject to legislative appropriations for the purpose." The question of these rebates or subventions is a difficult one. While they have certain theoretical merits in terms of fiscal federalism, they raise a number of practical problems from an administrative viewpoint. An informed study of this subject has yet to be made.

Turning to the second part of Mr. Banta's chapter, three extensions to use-value farmland assessment are considered: voluntary agreements, acquisition of development rights, and exclusive farm use zoning. With respect to voluntary agreements, Mr. Banta notes that the California and New York programs have not attracted a high rate of participation near urban areas. The same seems to be true in Hawaii, where landowners who are willing to commit their land to farm use for either a ten- or twenty-year contract period are eligible for assessment at one-half of use value, but with full rollback penalties plus ten percent interest for early withdrawal. The failure of such programs in this regard should probably not be taken too seriously. After all, it has been rather clearly established that use-value assessment programs cannot by themselves prevent urban sprawl. Mr. Banta perceptively notes that those programs characterized by voluntary agreements (contracts) are only technically different from other programs involving an application and deferred penalty. Perhaps the legal rights of the parties are more clearly established, although it is doubtful that this has much effect on the behavior of potential participants. In any case, the real merit of use-value assessment programs is that they offer an *opportunity* to the farmer whose land is threatened by high taxes due to nonfarm factors to continue his operations. That a large percentage of urban fringe farmers prefer, instead, to take a capital gain and either farm elsewhere or join the urban work force may or may not be an unfortunate situation, but does not mean that the opportunities offered by deferred taxation should necessarily be withdrawn.

The purchase of development rights, of course, is an effective means of ensuring farmland preservation. The high costs of acquisition, however, make it imperative that program objectives be clearly defined. Does, in fact, the state or local government perceive the continuation of farming on the affected lands to be in its long-run best interest? And is it worth the cost involved? Suffolk County, New York, appears to have answered these questions in the affirmative and is proceeding with a positive acquisition program as a result. In other instances, however, it may be that fee simple purchase of selected open space or farmland is in the better long-run interest of the community. Outright purchase may be little more expensive and, equally important, gives the government the flexibility of control over the long-range use of the land. In any case, purchase of either the development rights or fee simple interests in land provides a guarantee that tax expenditures will not have been in vain.

Mr. Banta raises certain problems with respect to the determination of use-values for lands for which development rights have been sold. The situation he describes in Burlington County, New Jersey, in which use-value assessments and farmland rentals differed highly from appraisals used to support development rights acquision, seems natural. The fact that use-value assessments would increase as the result of development rights acquisition implies that use values are currently understated. The key to a successful wedding of development rights acquisition and use-value assessment is to institute realistic use-value assessments *beforehand*. This tends to ensure that use-value assessments will, on the average, reflect the difference between full market values and acquisition costs. Use-value assessments, however, should continue to be based on independent criteria (e.g. soil productivity data), not on the difference between hypothetical market values and development rights costs. Realistic assessment also helps to ensure the success of the acquisition program itself.

With respect to exclusive farm use zoning, Mr. Banta notes that this is an inexpensive but politically difficult alternative: "In the long-run, however, it appears to be a necessary ingredient in a farmland protection program, at least at the urban fringe." This seems open to doubt since zoning land on the urban fringe for exclusive farm use opens a pandora's box of economic and political problems. It seems unlikely that zoning would perform better in this case than it has in other situations of intense development pressure. On the other hand, the exclusive farm use zoning of prime agricultural lands *beyond* the urban fringe, if properly carried out, appears to be operationally more feasible with a better chance of long-range success.

In conclusion, there is a need for state/provincial and local governments to make a hard, honest evaluation of their long-run land-use objectives. There is not an acute shortage of farmland, such that the loss of additional acres to urban development need be cause for grave concern. Instead, it is important to ensure that urbanization takes place in an orderly manner, with planned open space and recreational areas, and that farming as a whole be kept vigorous and profitable through appropriate national and sometimes state/provincial policies. In this scenario there is indeed room for well-formulated use-value farmland assessment programs. It should be realized, however, that such programs are (or should be) as much social as economic in their objectives and should be evaluated accordingly.

FIVE

The Politics of Tax Preference
JOHN BRIGHAM

Public policy is similar to fashion in that the statutes through which it is expressed go in and out of vogue. The policy examined here is a tax preference for farms and open space land. It was in vogue from 1957 to the early 1970s. During this period, over 40 states decided to tax certain property owners on the basis of assessed valuations representing the use rather than the market for their land. It is, without some compelling explanation, curious that these tax preferences swept the country in less than 20 years. Paradoxically, in an essay recently published as part of a major symposium on property taxation and the political system, it was proposed that "honesty forces us to admit that changes in the property tax system have been slow and not always in the direction recommended by those who have made impartial studies of the tax" (Fisher, p. 5). It is the thesis of this chapter that this particular change has been neither slow nor characterized by "impartial studies." The successful effort to introduce this policy change is the subject of this analysis.

When one examines conditions that influence the policy process at the state level, distinct facets of the process are difficult to discern because of the limited context in which each statute is passed. If we turn from the rationales to the conditions affecting policy, we find the states in a policy formation process in which de facto preferences for farmland, traditionally maintained along with many other extralegal tax preferences, were put into law. The change is best understood in terms of the movement for 100 percent valuation of property which was itself stimulated by the growth of state aid and the effort to equalize local expenditures. Although occasionally acknowledged to have a relationship (Hagman, p. 637), these factors have not been systematically put forth as explaining the statutory tax preferences. Since these preferences reflect unacknowledged shifts in tax policy to preserve the status quo, it is not surprising that they fail to achieve their purported land-use objectives (CEQ; Gustafson and Wallace; Gloudemans). In this regard the chapter suggests that the evaluation of public policy proposals look not at what is said about them (forward into the new world which they propose to create) but backward, as it were, to the changes that arise and interests that are brought to bear.

My analysis proceeds from the factors at the base of the policy process. Although infused with the ideological predispositions of citizens, in the form of a desire for suburban life, and of policy-makers, out of a desire to equalize local public expenditures, the phenomenon of material change establishes the basis for policy analysis since it presents the policy-maker with the conditions on which policy must be evaluated. In order to understand policy, it is not enough to be content with the ideological treatment it is given—the rationales which support it—but neither can these be ignored. Policy formation involves a response to conditions which is translated into law.

The status quo ante

The most obvious condition bearing on preferential treatment for farmland is the population shift to the suburbs and the extraordinary increase in the per capita consumption of land. This dramatic change in land use from the early 1950s to the mid 1970s is common knowledge. It transformed the landscape and altered the conditions under which people lived. Metropolitan growth during this period involved converting approximately 22,000 square miles of land from rural to urban uses (Brown). As Roberts pointed out in the first chapter, the rate increased from 30 acres of land for 1,000 new residents prior to the Second World War to 200 acres for the same number of new suburban residents in the last few years. The figures are even more dramatic for the land that went out of farming due to the encroachment of the city. Barlowe (1967, p. 92) reported the loss of 22 million acres to farming in the period from 1954 through 1959. Of this total, only one-third is reported to have gone into development and two-thirds of the land to have reverted to lesser uses due to marginal agricultural conditions as well as to off-farm opportunities for what has been termed "greater income satisfaction" (Barlowe, 1967, p. 93). The loss of farmland overall is related to the intensity with which the agricultural establishment has turned to agricultural preservation. The rising value of urban fringe land is central to the explanation of the rise of tax preference schemes.

Until relatively recently, a variety of de facto tax preferences existed for farms and homes (Bonbright; Gloudemans; Hagman; Netzer, 1968). Some stemmed directly from increases in land values at the urban fringe. The rise in market values led to a situation of nonmarket assessment, described as "assessment lag" (Paul, p. 26; Peterson). When land increases in value it comes to be assessed at a diminishing percentage of its market price because the assessment lags behind the market. Thus, the simple pace of reassessment, combined with rising value, leads to a nonmarket tax base. Clawson has pointed out that frequency of reassessment must be considered in determining the tax costs on suburban land in a rising market (p. 122). Engle has characterized as de facto discrimination the shift in the tax burden toward property with a slow rate of increase in value.

De facto preferences depend on the extralegal tradition of partial assessment. This key characteristic of property tax administration was dramatically laid out over 40 years ago in the influential treatise by James Bonbright, *The Valuation of Property*. He proposed that "one of the strangest characteristics of the property tax, and its most striking departure from the standard of market value, was the all but universal prevalence of percentage or partial valuation" (p. 497). Bonbright suggested that misinformed taxpayers associate a larger valuation with a larger tax and are not as contentious about the tax if they believe it is based on something less than their property is worth (p. 498). After analysis of 117 cities, John Shannon concluded that farmers and homeowners are "more frequently the beneficiaries not the victims—of extralegal assessment practices" (p. 30). Those whose farms were increasing in value not only received the benefits of "assessment lag" but, in addition, generally received favored treatment in the local political environment that has controlled the property tax until recently. Where the state relied on the property tax for part of its revenue, local assessors often sought to lighten their constituents' burden at the expense of the other localities (Bonbright, p. 498). This practice of partial valuation continued in the face of constitutional provisions for uniformity and equality in taxation and numerous court decisions that acknowledged the inequalities resulting from less than full valuation (*Fletcher Paper*, 1910; *Greene*, 1917; *Oglesby*, 1930). In fact, it is generally agreed to have increased with efforts to distribute state money in an inverse relationship to community wealth.

State involvement and equalization

Against this backdrop, and throughout the same period, another change operated to create the environment for statutory preferences. In addition to conversion at the urban fringe and increased land values in the context of partial assessment practices, state involvement in local funding with a redistributive intent was a crucial change that came to affect the politics of assessment. Just before tax preferences for agriculture began to sweep the country, a new state involvement in aid to localities became evident and continued throughout the period. The great increase of state intergovernmental transfers in the last 40 years is a dominant issue in the financing of local government (Maxwell and Aronson, p. 85). Since one function of this involvement was an effort to equalize state education aid, it produced a far greater sensitivity to local assessments than existed previously. Although litigation examining the state effort on the basis of equal protection did not come until the 1970s (*Serrano*, 1971; *Robinson*, 1972; *Milliken*, 1972; *San Antonio*, 1973), the major cases reveal the ongoing efforts to equalize state funding for education.

The disparities in the tax base between school districts began to be evident after the Second World War due in large measure to the constitutional foundations of equal protection as applied to education. Concern about equality in education led to efforts to ameliorate the differential. In the case of Texas, this was achieved through the Texas Minimum Foundation School Program which was similar to those in other states. This program operated under a formula for distributing school aid in proportion to need. By early 1970s, the program accounted for nearly half the total educational expenditures in Texas. At the same time that state involvement in local funding began to increase, the administration of the property tax began to be called into question. Both partial assessment and interdistrict disparities came under fire from the National Tax Association and state governments. This resulted in attention to the state equalization function and the way it was being administered. Commentators were generally optimistic about the future of tax equity while criticizing past practice (Lee; Welch; Yale). A real equalizing activity was deemed necessary in light of the increased reliance on state funding. This prepared the foundation for equality of assessment which in turn spawned statutory tax preferences.

Uniform valuation

New standards tended to rely on market value as a guide to equality in taxation. The reform movement has been described as assuming that state governments will be the prime movers in rehabilitating the property tax (Maxwell and Aronson, p. 6). Although legislative involvement was evident in many states (Donovan, pp. 53–54), this sensitive political issue was handled primarily by the courts. One observer characterized the series of decisions beginning in the late 1950s as a "nuclear bomb" on the 100-year tradition of fractional and inequitable assessments (Becker, p. 36). The trend since the late 1950s has been toward successful litigation that either ordered full market valuation or required uniformity within a taxing district.

The cases followed a pattern of attention to local inequalities in the form of review and to statewide or interdistrict inequalities in the form of equalization. Most cases have involved inequities within taxing districts and began to exert judicial supervision in spite of extralegal partial assessment (*Hammermill*, 1952; *Baldwin Construction*, 1954; *Bemis*, 1954; *Hamm*, 1959). Nevertheless, interdistrict concerns that reflect equalization in state aid have influenced the judicial remedies for intradistrict discrimination.

Prior to this movement, review had been nearly impossible unless an individual assessment exceeded market value (*Ingraham*, 1957). In *Hamm v. State*, the Minnesota Supreme Court held taxation of some property owners at a higher percentage of market value to be a denial of the equal protection clause of the Fourteenth Amendment of the federal constitution and a violation of the uniformity provision of the state constitution (1959:653).

In other decisions, courts began to order reassessment on the basis of statutory requirements (*Baldwin*, 1954; *Switz*, 1957; *Bettigole*, 1961; *Walter*, 1969; *Hellerstein*, 1975). That is, districts were ordered to reassess every property at the level of full or market value as the best standard of equality. New York produced an instructive decision from the Court of Appeals. Here, the requirement of statewide 100 percent valuation was influenced by decisions in other state courts. In *Hellerstein* v. *The Assessor of Islip*, Judge Wachtler noted the longstanding extralegal practice of partial valuation. On this basis, the challenge to partial assessment had been dismissed in the lower courts, although state law provided that "All real property in each assessing unit shall be assessed in the full value thereof" (1975:280). "Full valuation" had been required in the state since 1826, and a similar notion of "true value" went back to the Federal period. The custom of fractional assessment, however, was understood to be as old as the statute (Kilmer, p.

210), and until he reached the Court of Appeals, Hellerstein had failed in his challenge. Wachtler, however, ruled that decisions of the New York courts which accommodated the practice of partial assessment as recently as 1965 would have to be overturned in conformity with current concern for equity of the property tax. He argued that there was no uniformity in assessments and that Boards of Equalization had failed to deal with statewide disparities. With this decision, which referred directly to the cases alredy mentioned, New York became one of the last states to require 100 percent valuation. By the time the decision had been made, a trend that associated state involvement in local financing and efforts to equalize public expenditures had become evident. It is not inconsequential that New York was one of the last of the northeast states to provide statutory tax preferences for farmland and open space.

In the period under study, these orders had a greater impact than earlier decisions reasserting a statutory obligation. In Massachusetts, during the five-year prior to the decision in *Bettigole*, 32 of 351 communities revaluated. In a similar period following the decision, 177 of the communities in the state revaluated (Paul). Enforcement of the local obligation to assess property for tax purposes on the basis of full value was intensified when litigation began to reflect inequities between communities. It is most clearly represented by suits filed by communities against state administrations (*Sudbury*, 1973; see also *Louisville*, 1966). The claims were based on the loss of state aid which would go to towns that underassessed themselves as a result of equalization principles. In Massachusetts, where full valuation had been required in various cities for up to 14 years, the *Sudbury* decision ordered the State Tax Commission to enforce the constitutional requirement of full-value assessment for all 351 cities and towns in the state.

State supervision over the property valuation process, whether involving inter- or intra-district discrimination, created a new environment for the property tax. It emphasized uniformity of assessment in practice consistent with statutory requirements and brought about an obligation for periodic assessment that was particularly burdensome to owners of property that was increasing in value. The response was for statutory preferences, the most widespread being that for agricutural land.

Valuation and legislation

The association of preferences with uniform valuation ties them to agricultural interests in a way that is not generally acknowledged. Unlike the land conversion and open space rationales publicly identified with the schemes, this factor antedates the environmental movement and thus provides an important independent variable that covers the entire period of legislative change.

In New Jersey, uniform valuation and statutory tax preferences for farmland are associated with Olivia Switz, who forced court-ordered 100 percent valuation in 1957. Following her initial success, the property tax for farmers had gone from $5 below the per capita average to $222 above the average (Hagman, p. 638). The state legislature then passed a special preference for farmers. Ms. Switz had to return to court four years later to protect her ruling. She was successful again (*Switz*, 1962), but continued interest in New Jersey and elsewhere in providing preferences for farmers did not cease. In Maryland, the imposition of full-value assessment in Montgomery County provided the impetus for the first use-value preference for agricultural land (Hagman, pp. 637–638). A study of the preference in Connecticut reveals that the farmer has little or nothing to worry about until his land is threatened with market valuation (CEQ). Prima facie evidence thus exists for an interpretation showing that these preferences have been sought where large landowners were faced with reassessment and the loss of de facto tax preferences.

Maryland, Connecticut and New Jersey were among the first to pass the preference legislation. Passage by Florida in 1959 followed two years later with state-mandated full-value assessment (Florida Stat. §193.11, 1961). Here, however, it had been preceded by county action and a strong push from the governor's office that made the new standard of assessment more realistic (Donovan, pp. 53–54).

The comparison of tax preferences with the movement to 100 percent valuation is difficult because of the different traditions in each state as to the status of the de facto preferences and the variety of struggles that preceded passage of the de jure preference. The Massachusetts Farm Bureau, for instance, began its efforts to change the law on agricultural land assessment in 1958 following the change undertaken by Maryland. (This was prior to the first 100 percent valuation case in Massachusetts.) The law was defeated in that year, and in 1966, 1967 and 1968. The basis of support in the early years was the agricultural community as the law emerged from the district of the state's land grant college and agricultural extension headquarters. It was ultimately necessary in 1969 to pass a constitutional amendment making tax preferences legal. In 1973, relatively late in the movement for statutory preferences, the Massachusetts House of Representatives voted 218 to 10 in favor with 11 members not voting. The size of the vote is remarkable in itself and suggests the legislation had become overwhelmingly popular and essentially without identifiable opposition. Although a year later the legislature defeated a statutory proposal which would have implemented earlier court decisions requiring full-value assessment, the courts of the state instructed the executive branch to enforce full-value assessment within the year.

The California Land Conservation Act of 1965 discussed by John Banta in Chapter 4 (The Williamson Act) was preceded by earlier preference legislation in that state. Although deemed unsuccessful, the Agricultural Assessment Law of 1957 provided tax preferences on the basis of zoning and represented early state policy that sought to induce farmers to retain their land. The intense pressure of land conversion joined with a strong agricultural industry explains experimentation with preferences that preceded the Maryland case, the earliest generally recorded. In California, property tax reform came in the mid-1960s (Revenue and Tax Code. Sec. 401, 1966) coinciding with the Williamson Act, but subsequent to the earlier statutory preferences. The reforms in California, like those in some other states, provided for a fixed percentage of market value as the basis for assessment.

States required to revaluate early, like Florida, or fearing the mandate, as in Maryland, were among the first to pass preference legislation. Where a state was surprised by the Court decision, as in Connecticut, New Jersey and Minnesota, it generally took about six or seven years until the legislation was passed. New York passed the preference prior to being forced to 100 percent valuation. But in this case, by the time the decision had been handed down in 1975, the movement for some farm preference had enough momentum to stimulate legislation. It thus transcended the original factors by which it can most readily be explained.

Kansas, which by June of 1978 was one of the handful of states that had yet to pass the preference legislation, provides an excellent example of the effects of the de facto tax structure on legislation. Prior to consideration of the special preference for agriculture, the state had mandated 30 percent valuation, but only oil and gas and state-assessed public service companies had been brought to that level. The assessments on homes and farms had not yet been implemented. As a result, an impact study of use-value assessment considered that "If all property is reappraised and if use value appraisal is implemented, a partial shift to residential and other real property (excluding agricultural land) will occur. If use value is implemented without reappraisal of other property the shift will be to agricultural land from other property including residential" (KSU, p. 13). This is a consequence of farmland assessment being at less than 30 percent with the shift to use value bringing them to an even higher level of assessment. As a result, Kansas has been hesitant to go the the use-value scheme prior to any compulsion to go to a stipulated percentage of market value in other areas.

The introduction of statutory tax preferences has thus been characterized by interest group success in maintaining the status quo—not because the farms which would be affected were paying more than their share but because they were paying less. The movement to uniform assessments generally at full market value is the most important factor during the period under study in explaining this interest group activity. Some demographic variables throughout the states assessed in terms of the rate at which the states accepted the scheme are examined in the following section.

Other determinants of preference legislation

The point at which a state passed the preference for farmland during the last 20 years is also directly related to the significance of agriculture, the per capita property tax, and the amount of urban growth in the state. These are conditions that increase legislative sensitivity to this policy throughout the period and indicate the basis for response to changes in valuation policy. They all heighten the significance of full valuation when it comes and may explain why this policy change is not directly related to a state's propensity to be innovative in other policy-making.

The first ten states to pass the legislation* are characterized by strong agricultural interests and more dramatic urban population growth than the rest of the nation. They average $760 million in the value of farm products in 1964, at the center of the period under study, against a national average per state of $705 million (Census, 1977:679). They average 26.35 percent in change in the metropolitan population between 1960 and 1970 against the national average of 17.10 percent (Census, 1977:17). While the range in growth rate is small between these states, the range in agricultural wealth produced is large enough to undercut further the significance of this figure. In these cases, growth, by pushing up land prices, activated the agricultural community to maintain use-value assessment. Of the eight states which had not passed a preference by 1975†, the average value of farm production for 1964 was $673 million and the change in metropolitan population was 13.37 percent from 1960 to 1970. The population shift is lower than that of the nation and half what it is for the states instituting the preference early.

*Maryland, Florida, Indiana, Hawaii, Alaska, Connecticut, Oregon, New Jersey, California, Pennsylvania (CEQ, 1976).

†Alabama, Georgia, Kansas, Louisiana, Mississippi, Tennessee, West Virginia, Wisconsin (CEQ, 1976).

In examination of the characteristics of the property tax itself, the most significant factor is that none of the nine states with the lowest per capita property tax in the country* passed legislation before the last decade of the period under study, and five of the nine are among the eight states that didn't pass the legislation at all. In addition, four of the first seven states to pass the legislation rank in the top ten in reliance on the property tax (Phares, p. 67). This factor in itself explains little, however, since three of the others in the top ten never passed the legislation, and the other three did relatively late. This further elevates the importance of assessment policy. Of the states where taxes on farm property were the highest as a percentage of farm income in 1974†, only one passed the preferences early and the others were noticeably late.

Political scientists have been experimenting with measures of innovation in the states (Gray; Savage; Walker) or the propensity to make policy changes. Although they have been subject to various criticisms, the more recent contributions suggest an improvement, and in any case they provide a rough test of the influence of state political culture on the adoption of legislation. Savage used 181 policy measures to develop a scale of innovativeness for three periods in American history (pp. 216–217). These scales provide a basis for examining what might be an important dimension of policy change by states. His scales show some consistency over time with a general trend toward the adoption of novel policies with increasing speed, a function he attributed to nationalizing forces. Since the items on which Savage built his scale for 1930 to 1970 involve considerable regulation, taxation and state aid, his scale may reflect a progressive or liberal inclination. Because tax preferences have generally been billed as this sort of measure, a general limitation in the scale does not necessarily detract from its utility here. Of the ten states passing the preference in its first decade, only three (California, Oregon

*Alabama, Arkansas, Kentucky, Louisiana, Mississippi, North Carolina, South Carolina, Tennessee, West Virginia (Netzer, 1966:90–91).

†Massachusetts, 19.8; New Jersey 16.6; New York 17.5; Rhode Island, 18.9 (USDA, 1976).

and Indiana) ranked at the top of the innovativeness scale established for the period 1930–1970. On the other hand, of the eight states that had not legislated a preference by 1975, three fell into the lowest decile on the measure of innovativeness (Alabama, Georgia and Mississippi). Although none of the "most innovative" states failed to pass the preference and none of the "least innovative" passed it early, the generally low correlation of tax preference legislation with measures of innovativeness for individual states lends support to the thesis that this legislation does not represent meaningful innovation but is rather a policy that preserves the status quo.

Research and rationales

As a social phenomenon, the rise and implementation of use-value assessment has at its core the protection of agricultural interests. In this context, it has been viewed as both a necessary measure to preserve the economic viability of farming (Engel) and as a classic case of special interest legislation (Stocker, 1976). Early commentary was not entirely favorable to the legislation as a means of keeping land in farming (Stocker, 1961). Yet in nearly every state that passed the legislation, research describing its virtues was carried out by the extension service and experiment station network. This research proposed that use-value taxation was preferable to public ownership and other forms of regulation (Engel). Conducted first by the Department of Agriculture and then by the Extension Service in each state, the research was influential in getting the legislation through. The tax preference was linked to public concern for open space only in the more recent studies (Nelson, p. 215).

While all of the 42 states that had passed preference legislation by 1975 included agriculture as a basis for the assessment, only 18 included forests. Even fewer, 13, brought open space into the protection, and only eight states had offered the tax benefit to recreation areas. These characteristics of the various statutes confirm that the preferences are primarily farmers' statutes. Looking at preferences for open space alone, there is no evidence to indicate that the legislation changed its characteristics as the environmental movement gained momentum during the period under study. Hawaii, with one of the earliest bills, included the open space provision in 1961, and the other 12 states including this provision are distributed throughout the entire period in which this preference was institutionalized.

To say that the legislation is farmers' legislation rather than environmental has in the past been enough to indicate the extent to which the legislation also served the interests of speculators. In the political activity surrounding passage of these statutes, speculators have not been obvious, but there is no doubt that speculators have benefited considerably from statutes which could not have been passed with their interests in the forefront. One way to gauge the interest of speculators is to compare the bills which are particularly advantageous to them and those which are less favorable. In 1975, only eight states* had pure preferential assessment without additional eligibility requirements other than that the land be in a stipulated use. Of these, only Arizona and Indiana combined agriculture with considerable urban growth. The political culture of Arizona has been particularly susceptible to the interests of speculators (see Brown and Roberts). The largest portion of the states fall under a scheme of deferred taxation with limits on eligible use. It is the 26 states in this category that represent the dominant assessment scheme. Although such schemes are not as favorable to speculators as pure preferential assessment, they constitute tax and land use systems far more favorable to large land holders than most other options. This is evident especially when one turns to the most restrictive formulations. Five states have provisions to keep land undeveloped for an average of ten years by means of restrictive agreements. Yet it is from the experience of one of these states, California, that we learn of the extensive use of the preferences for the benefit of speculators and large land holders (Fellmeth; Wagenseil). The legislation is best characterized in terms of farmer and speculator interests rather than those of environmentalists or planners.

The early studies coming from the agricultural research

*Arkansas, Arizona, Idaho, Indiana, Iowa, New Mexico, North Dakota, and Oklahoma.

establishment did not assess the influence that the tax prefer-ence was likely to have on the behavior of farmers. This did not come until the legislation was passed, and it reflects an environmental rather than an agricultural concern. Farmers knew it wasn't going to hurt them, but the preferences were supported by environmentalists and planners who seem to have accepted the agricultural research which presented the statutes as a means of preserving farmland. These groups did not add their research capacity to assessment of the preferences until after they had been put into law. Disenchantment with the mechanism has begun to be evident from these quarters. In a recent evaluation for the Council on Environmental Quality (1976), the authors propose, in one case study, that "while differential taxation was promoted under the slogan, 'Save Our Open Space,' preservation of open space was not the real concern of the farm and timber interests who were the major forces seeking the legislation" (p. 236).

The involvement of land use interests is of a different sort than that emerging from the agricultural community. Farm interests provided the impetus and environmentalists joined in, perhaps naively. Environmentalism, although it generated considerable attention two-thirds of the way through the period of legislation, was markedly immature compared to the longstanding and highly institutionalized interests of agri-culture. The environmental movement was just beginning to be a major political force by 1970 when it was confronted with the already activated preferential assessment bandwagon com-ing out of the agricultural research establishment. Environ-mentalists may have been overly eager for a victory and thus willing to accept the intuitively correct if superficial claims for a relationship between tax preferences and the preservation of farmland and open space.

Once environmentally conscious research began to probe these tax preferences, the results were far more critical than had previously been available. In particular, conservation groups evaluated who benefited and the extent that the preferences affected decisions about whether or not to stay in farming. Some of the most critical studies concerned one of the most restrictive statutes. Wagenseil pointed out that "only 6.4% of the land under contract in the Williamson Act was located less than three miles from cities" (p. 190). According to his assessment, those close to the cities and hence the most likely to face development pressure chose not to tie up their land. He concluded that there was no reason to believe that use-value assessment "will be the major tool in the curbing of urban scatteration and the preservation of open space" (p. 192). A Nader study a few years later confirmed these findings for California, reporting that within a three-mile radius of the cities, less than 5 percent of the land was under contract (Fellmeth, p. 41). Large-scale developers and landholders can plan decades ahead and have thus been able to "use the Act for land which is not scheduled for immediate development" (CEQ, p. 293). In an evaluation of how the statute has been employed for the town of Amherst, Massachusetts, the propensity for the greatest use to be made of the preference by large landholders is confirmed. Here, out of a total of $1,788,130 in property that has been placed under the preference, sixteen parcels of land with a market value of $604,770 belong to one institution, Amherst College. Under the state's use-value provision, it is being taxed at a value of $54,540.

Farmers have been the main beneficiaries of tax preferences that have kept their property taxes relatively constant per $100 of the market value of their land since 1940 (USDA, 1975a, p. 1). The research on which the legislation was based advanced these interests and suggested the contribution the statutes would make to the preservation of farmland. Not until the legislation had swept the country did it become clear that it failed to accomplish its purported land-use objectives.

Conclusion

The limitations of tax preferences in fulfilling land-use objectives have stimulated a new set of possible rationales for the statutes which rely on continued optimism. Tax preferences are seen as "At best . . . a holding action which may buy some time to develop alternative public land resource management methods" (Engel, p. 222). The preference schemes, however, seem to be no more a necessary first step to preserving farms and open space in the face of a rising market for land than price supports for agricultural products have been. In both cases, marginal improvement in the welfare of the farmer against the extraordinary advantage that may be gained by turning farmland into house sites presents a situation with minimal likelihood of realizing land-use objectives.

In policy analysis one is alerted to the limits of ideological rationales when the goals most generally stated seem not to be served and when all the research that was done in support of the schemes fails to show what subsequently turns out to be "obvious" limitations (see Chapter 3 by Coughlin; and Chapter 2 by Ladd). The pervasiveness of tax preference legislation indicates the need for a material explanation which is sensitive to ideological rationales for their capacity to gain acceptance in the political sphere. The position taken here, and the data presented, call attention to the dependence of policy change on material factors not evident in the ideological positions. It has been argued that if public policy on land is to achieve any goal, policy-makers must comprehend and confront the material determinants of legal change.

While the goals on which policy is made may be subject to cost-benefit analysis (Ladd), the legislative rationales through which the goals of policy may be discerned are not always reflective of the conditions that stand as the best indicator of why a policy was passed. If the conditions were adequately revealed, however, the results of legislation would be less surprising. Where the preservation of a traditional arrangement with regard to taxes on land is the condition, it might have been predicted that the alteration of land use would have been minimal.

By giving special favors to farmers in the form of a tax break, paradoxically due in part to rising concern for equity in tax policy, a link between land policy legislation and the public interest may yet be emerging. In one sense, at least, widespread implementation of these tax preferences has laid the foundation for meaningful exercise of land-use controls. It has created a group who benefit at public expense on the basis of a desire to preserve farmland and open space from development. Since other devices will obviously be necessary to reach the land-use goal (Banta; Stocker, 1976), the existence of past benefits may be grounds for future, perhaps unwilling, sacrifice. As Wagenseil suggested, the next steps will be a real test of "how highly open spaces are valued in our society" (p. 192). We are more capable of passing the test when we recognize that regulation or public purchase of development rights which actually preserve land will result in low taxes that are not preferential. The legislation discussed in this chapter is preferential because it gives a tax break but does not represent public policy on the use of land.

The success of the preferences in being passed may well be the key to their failure to contribute to the goals that were generally expected. If meaningful regulation of land use were to become national public policy, the ideas of preserving farmland and open space through preferential taxation would become an artifact of the failure of interest group pluralism to carry out a planning function.

References

Banta, John. "Operationalizing Differential Assessment for Agricultural Land." (Chapter 4, this volume.)

Barlowe, Raleigh. "Taxation of Agriculture" in Richard W. Lindholm, *Property Taxation: USA*. Madison: University of Wisconsin Press, 1967.

Barlowe, Raleigh, and Alter, Theodore R. (1976) "Use-Value Assessment of Farm and Open Space Lands." Experiment Station Research Report no. 308. East Lansing: Michigan State University, 1976.

Becker, Arthur P. "Property Tax Problems Confronting State and Local Governments." In H. T. Johnson, ed., *State and Local Tax Problems*. Knoxville: University of Tennessee Press, 1969.

Bonbright, James C. *The Valuation of Property*. New York: McGraw-Hill, 1937.

Brown, H. James, and Roberts, Neal A. *Land Into Cities: The Land Market on the Urban Fringe*. Cambridge, Mass.: Harvard University Graduate School of Design, 1978.

Clawson, Marion. *Suburban Land Conversion in the United States*. Baltimore: Johns Hopkins University Press, 1971.

Coughlin, Robert E. "The Magnitude of the Tax Savings and Its Effect on the Land Market." (Chapter 3, this volume.)

Council on Environmental Quality. *Untaxing Open Space: An Evaluation of Differential Assessment of Farms and Open Space*. Washington, D.C.: GPO, 1976.

Donovan, C. H. "Recent Developments in Property Taxation in Florida." In Harry T. Johnson, ed. *State and Local Tax Problems.* Knoxville: University of Tennessee Press, 1969.

Engel, N. Eugene. "Political and Economic Forces Behind State and Local Approaches to Retain Prime Land." USDA Seminar on Retention of Prime Lands. Washington, D.C.: GPO, 1975.

Engle, Robert F. "De Facto Discrimination in Residential Assessments: Boston." *National Tax Journal* 28 1975): 445–451.

Fellmeth, Robert C. *The Politics of Land: Nader Study Group Report on California.* New York: Grossman Publishers, 1973.

Fisher, Glenn W. "Property Taxation and the Political System." In Arthur D. Lynn, Jr., ed., *Property Taxation, Land Use and Public Policy.* Madison: University of Wisconsin Press, 1976.

Gloudemans, Robert T. *Use Value Farmland Assessment: Theory, Practice, Impact.* Chicago: International Association of Assessing Officers, 1974.

Gray, Virginia. "Innovation in the States: A Diffusion Study." *American Political Science Review* 62 (1973): 1174–1185.

Gustafson, Greg C., and Wallace, L. T. (1975) "Differential Assessment as Land Use Policy: The California Case." *Journal of American Institute of Planners* 41 (1975): 379–389.

Hady, Thomas F., and Sibold, Ann G. *State Programs for the Assessment of Farm and Open Space Land.* Economic Research Service, USDA. Washington, D.C.: GPO, 1974.

Hagman, Donald G. "Open Space Planning and Property Taxation—Some Suggestions." In *Wisconsin Law Review* 628, 1964.

Harvard Law Review. "Inequality in Property Tax Assessments: New Cure for an Old Ill." *Harvard Law Review* 75 (1962): 1374.

Hastings Law Journal. (1968) "Land: Unraveling the Urban Fringe." *Hastings Law Journal* 19 (1968): 421–427.

Heller, T. C. "Theory of Property Taxation and Land Use Restriction." *Wisconsin Law Review* (1974): 751–800.

Johnson, Harry T., ed. *State and Local Tax Problems.* Knoxville: University of Tennessee Press, 1969.

Kansas State University "Use Value Appraisal Impact Study." Manhattan, Kansas: Department of Economics Cooperative Extension Service, 1977.

Kilmer, Robert F. "Legal Requirements for Equality in Tax Assessment." *Albany Law Review* 25 (1961): 203.

Kolesar, John, and Scholl, Jaye. "Misplaced Hope, Misspent Millions: A Report on Farmland Assessments in NJ." Princeton, N.J.: Center for Analysis of Public Issues, 1977.

Ladd, Helen. "The Tax Policy Considerations Underlying Preferential Tax Treatment of Open-Space and Agricultural Lands," (Chapter 2, this volume.)

Lee, Eugene. "State Equalization of Local Assessments." *National Tax Journal* 6 (1953): 176.

Lynn, Arthur D., Jr., ed., *Property Taxation Land Use and Public Policy.* Madison: University of Wisconsin Press, 1976.

Maxwell, James A., and Aronson, J. Richard, *Financing State and Local Governments*, 3rd Ed. Washington, D.C.: The Brookings Institute, 1977.

Nelson, B. E. "Differential Assessment of Agricultural Land in Kansas." *Kansas Law Review* 25 (1977): 215–45.

Netzer, Dick. *Economics of the Property Tax.* Washington, D.C.: The Brookings Institution, 1966.

————. "Impact of the Property Tax—Effect on Housing, Urban Land Use, Local Government Finance." National Committee on Urban Problems, Research Department, No. 1. Washington, D.C.: GPO, 1968.

Paul, Diane. *The Politics of the Property Tax.* Lexington, Mass.: Lexington Books, 1975.

Peterson, George E., ed. *Property Tax Reform.* Washington, D.C.: The Urban Institute, 1973.

Phares, Donald. *State-Local Tax Equity.* Lexington, Mass.: Lexington Books, 1973.

Roberts, Neal A. "The Big Giveaway Called Differential Assessment." (Chapter 1, this volume.)

Savage, Robert L. "Policy Innovativeness as a Trait of American States." *The Journal of Politics* 40 (1978): 212–224.

Shannon, John. "The Property Tax: Reform or Relief?" in George Peterson, ed. *Property Tax Reform.* Washington, D.C.: The Urban Institute, 1973.

Stocker, Frederick D (1961a) "How Should We Tax Farmland in the Rural Urban Fringe?" USDA, Economic Research Service. Washington, D.C.: GPO.

————. (1961b) "Some Problems in Assessing Farmland in the Rural Urban Fringe." USDA, Economic Research Service. Washington, D.C.: GPO.

————. "Property Taxation, Land Use, and Rationality in Urban Growth Policy," in Lynn, op. cit., 1976.

Census, U.S. Bureau of *Statistical Abstract of the United States.* Washington, D.C.: GPO, 1977.

USDA (1975a) Economic Research Service. "Farm Real Estate Taxes— Recent Trends and Developments." Washington, D.C.: GPO, 1975.

USDA (1975b) Perspectives on Prime Lands: Background Papers for a Seminar on Retention of Prime Lands, July 16–17, 1975. Washington, D.C.: GPO, 1975.

Wagenseil, Harris. "Property Taxation of Agricultural and Open Space Land," *Harvard Journal on Legislation* 8 (1970): 158–196.

Walker, Jack L. "The Diffusion of Innovation Among the American States."
American Political Science Review 63 (1969): 880–889.

Welsh, Ronald B. "Intercountry Equalization in California." *National Tax
Journal* 10 (1957): 57–66; 148–157.

Yale Law Journal. "Tax Assessment of Real Property." *Yale Law Journal* 68
(1958): 335.

Cases cited

Baldwin Construction v. Essex County 108 A2d 598 (1954).
Bemis Brothers v. Claremont 102 A2d 512 (1954).
Bettigole v. Assessors 178 NE 2d 10 (1961).
Fletcher Paper v. City of Alpena 125 NW 405 (1910).
Greene v. Louisville 244 US 499 (1917).
Hamm v. State 95 NW 2d 649 (1959).
Hammermill Paper Co. v. Erie 92 A 2d 422 (1952).
Hellerstein v. Assessor of Islip 332 NE 2d 279 (1975).
Ingraham v. Town of Bristol 132 A 2d 563 (1957).
Louisville and Nashville Railway v. State Bd. of Equalization 249 FSupp. 894
(1966).
Milliken v. Green 203 NW 2d 457 (1972).
Oglesby v. Chandler 288 P 1034 (1930).
Robinson v. Cahill 287 A 2d 187 (1972).
Russman v. Luckett 391 SW 2d 694 (1965).
San Antonio v. Rodriquez 411 US 1 (1973).
Serrano v. Priest 487 P 2d 1241 (1971).
Sudbury v. Commissioner of Corporations and Taxation 321 NE 2d 641
(1973).
Switz v. Kingsley 182 A 2d 841 (1962).
Switz v. Middletown 130 A 2d 15 (1957).
Walter v. Schuler 176 So 2d 81 (1965).

SIX

Conclusion: How Ineffective Policies Gain Widespread Acceptance

DAVID E. DOWALL

Preferential tax assessment (PTA) of agricultural land has gained currency over the past 20 years. Beginning with Maryland in 1956, over 40 states have enacted legislation allowing for the preferential taxation of agricultural land. Recently, policy analysts have suggested that such schemes are ineffective and often work at cross-purposes with local and metropolitan land-use policies.[1] Despite growing criticism, preferential taxation of farmland continues to enjoy widespread use. This chapter summarizes criticisms outlined previously and suggests that preferential land taxation is best understood as a tax subsidy program, not as a device for achieving land policy goals.

Any effort to evaluate policy must first begin by defining the appropriate domain of evaluation.[2] Preferential taxation is an element of some broad policy space called land policy. Land policy is concerned with how land is used. It is based on judgments about the desirable types of land use, the location and intensity of land uses, and the "balance" among land uses within a geographic area. In the present context, land policy is concerned with the location and amount of agricultural land contained in a political subdivision. If agricultural land is being converted to urban or suburban uses at a rate thought to be too rapid, or if prime as opposed to marginal land is being converted, then land policy considerations may dictate that the character of such land conversion be altered to achieve desired objectives.

Such market intervention, economists argue, may not be justified on efficiency or equity grounds.[3] Efficiency considerations dictate that land utilization occur in a manner to ensure that the level of returns (holding constant for risk and talent) is equal between uses. Conditions of perfect competition, particularly mobility of factors of production, are important for ensuring that the marginal conditions of land use are maintained. Under conditions of perfect competition, PTA merely distorts the market and, as a consequence, misallocation of resources may occur. On the other hand, if there is evidence that significant externalities exist, economic policy considerations would imply that some form of intervention is warranted. For example, if agricultural land uses generate benefits such as open space and improved air quality that are not registered in market transactions, agricultural production perhaps occurs at a level below that desired by society because the benefits to society are not priced—the farmers do not receive payment for providing open space or clean air. Similarly, if the spatial development of cities is inefficient (e.g., low density urban sprawl) and assuming that the preservation of agricultural lands will reduce the amount of sprawl development, then as before it may be desirable to maintain agricultural land. Neither of these two cases has been systematically explored.[4] Instead, the rationale for preferential taxation of agricultural land is based on a set of preconceived notions about farmland conversion, and benefits derived from its control.

The farmers' plight

Preferential taxation developed because it was thought that farmers were being "taxed out of business" as urban areas expanded into the hinterland. As land ripened, the market value for it increased.[5] Market values are fundamentally different from use value in that they incorporate potential use value as well as current value. As the potential for land to convert from agricultural to urban uses becomes eminent, property assessors, if operating under the edict of market value assessment, will upwardly reassess farm land. This, it is argued, forces farmers to sell out to developers. Furthermore, advocates of PTA suggest that such sales are premature. On the other hand, where land is taxed at use value (a value reflecting *only* current use, farming), tax pressures are reduced significantly and land conversion is avoided.

Pressures for legislative change were and continue to be based on arguments that the preferential assessment of agricultural land will alleviate the pressures for farmland conversion. Preferential taxation is achieved by assessing farmland at use and not at current market value. As an element of land policy, it appears that PTA's popularity stems from its purported ability to maintain agricultural land in rapidly suburbanizing areas. Another alleged policy outcome is that it effectively subsidizes farmers and thus preserve agrarian values. This point is the key to understanding why such a patently decrepit policy instrument is so widely used. Preferential taxation is popular with state legislators because it supports agriculture and seems to be environmentally desirable and, in most situations, political and economic costs are widely dispersed. This broader perspective seems more capable of explaining the widespread use of preferential taxation of land. Having now set the stage for agricultural use-value taxation, let us turn to the four chapters. Each adds a bit more fuel to the ever growing fire criticizing the ability of PTA to achieve land policy objectives.

Preferential taxation of farm land as a tax policy[6]

A convincing argument for the widespread occurrence of PTA is that it involves neither direct expenditure for the preservation of farmland nor the postponing of suburban land conversion. To the noneconomist it looks like a free lunch. As Ladd's chapter suggests, such a naive view is incorrect. From a public finance and tax policy perspective, Ladd evaluates the use of preferential taxation of farmland.

Ladd argues that it would be more consonant with the principles of tax neutrality merely to defer taxes and not subsidize farmers. That is, instead of reducing tax liabilities through the use of preferential taxation, it is more neutral (not market distorting) to defer the payment of property taxes. When the farmland is sold, all back (deferred) taxes would be paid along with accrued interest. Such neutrality, however, may not be warranted. Ladd ignores possible land market externalities which may require the use of tax devices to improve economic efficiency.

If substantial externalities exist, it may be desirable to subsidize farming. Externalities such as nonpriced benefits of having land in agriculture, urban sprawl due to the under-pricing of transportation and housing, irreversibility of land uses, and the rising spectre of food shortages in the future may suggest that farming, because of its unpriced or long-term benefits, should be subsidized through tax concessions. If such externalities warrant government intervention, then the domain of policy evaluation expands to consider the ability of PTA to achieve a wider set of concerns.

Fundamentally, the role of government intervention is to alter the behavior of economic agents to promote market objectives of efficiency and equity. PTA, like any other tool of public policy, can be evaluated on such grounds. For a variety of reasons, Ladd suggests that preferential tax assessment is a "blunt" instrument for implementing land policy. For ex-example, she suggests that PTA generates only marginal changes in the supply of agricultural land while incurring costs to subsidizing taxpayers. California's Williamson Act, by restricting the benefits of taxation preference to those property owners who are willing to sign contracts for periods of ten years, affects only marginal decision-makers. Others have argued that in California, Williamson benefits flow to large corporate farming interests holding land that is not likely to be converted to urban use.

A second example of why PTA is only marginally effective is that it is not aimed at land planning objectives. There is absolutely no guarantee that PTA participants hold land in areas that are desirable for long-term agricultural use. Widespread participation in a statewide program does not guarantee that urban sprawl will be contained. Potentially, however, PTA could be very useful as an adjunct to community land-use control programs.

PTA, like any tax abatement program, confers benefits on recipients. The major evaluation question is whether the recipients are the right ones (farmers vs. speculators) and whether their behavior is altered in an appropriate fashion. Experience suggests that the recipients have been the wrong ones and that their behavior is not altered or that such behavioral changes are insignificant or irrelevant. Furthermore, preferential tax programs generate public sector costs that need to be evaluated in a manner similar to any other program of public expenditure. Tax incentive schemes involve costs. The major policy question is whether PTA yields results that are more cost effective than other forms of land use control. While Chapter 2 concludes that the direct purchase of farm land or the purchase of development rights may be more efficient than PTA, little evidence is available to support such an argument.

Up to this point, we have concerned ourselves with the efficiency aspects of PTA. Equity issues loom large. Use-value assessment of agricultural land is justified from an equity perspective if such a policy distributes burdens according to benefits or ability-to-pay principles. Chapter 2 outlines the equity issues associated with use valuation and taxation. Defining taxable income is a substantial problem that needs to be solved before we can prescribe the correct policy. Relying on ability-to-pay definition of tax equity, and defining income as annual income plus changes in net worth during the tax year,[7] Ladd concludes that use-value taxation is an inappropriate means to relieve the tax burden of farmers. "Even if restrictions are employed to limit relief under such programs to bona fide farmers, the relief is still not well targeted to those farmers with low ability to pay. Under use-value taxation schemes, the farmers with low ability to pay. Under use value taxation schemes, the farmers who benefit the most are those owning rapidly appreciating land. This means that farmers receiving the most benefits from use-value assessment are those farmers least in need in terms of ability to pay, correctly defined."[8]

A secondary equity question rests on the distinction between farmland and farmers' land at the urbanizing fringes of metropolitan areas. Tax policy directed towards landowners as a group will not necessarily aid farmers. Syndicates, investors, land speculators, etc., could benefit from tax concessions. Equity considerations might be met better by applying the familiar policy of tax circuit breakers to farmers. Transitional equity issues will plague any effort to comprehensively restructure tax policy. Movement from use-value to market value assessment will generate horizontal inequities. To avoid transitional problems tax policy changes should occur over a substantial time period, thus giving full warning to landowners and allowing ample time for market adjustment.

Chapter 2 addresses the administrative feasibility of use-value assessment. Evidence suggests that market value assessment has a slight administrative advantage over use-value approaches. Assessors cannot rely on comparables to reassess property since sales, particularly in urbanizing areas, reflect both use and market values. The inability to rely on sales transactions forces assessors to use the capitalization-of-income approach.[9] Begging the significant question of what rate of capitalization to use, the assessor is confronted with the problem of measuring farm income. In situations where annual income reporting is not feasible, assessors are forced to rely on the concept of "prudent management." Such an approach rewards inefficient farm operation. Most states, in attempting to avoid these problems, rely on productivity rates (e.g. soil classes) to value land. While administratively appealing, such crude approaches leave much to be desired since they ignore transportation differentials, topography, investments in drainage and cultivation practices particular to local areas.

Chapter 2 covered a substantial body of material relating preferential tax policy of land to the more general issues of tax neutrality, equity, efficiency, administrative feasibility and consonance with the principles of fiscal federalism. The evaluation as defined is excellent but its focus is narrow. Larger, more fundamental questions still remain.

The size of the tax saving and its influence on land markets (Chapter 3)[10]

Tax payments are derived from tax rates and assessed values. Together they determine the tax bill. Preferential assessments reduce the level of assessment for those land uses receiving special treatment. A smaller assessment with a constant tax rate yields a lower tax bill. Tax rates are determined by dividing the needed public funds by the total assessed value of property (regardless of the assessment procedures) located in the tax jurisdiction. Tax rates rise in response to either increased demands on the level of public services or reductions in aggregate property wealth. Differential assessment is promulgated to reduce the tax pressures generated by raising tax rates. Tax rates rise in areas experiencing urbanization where the demand for services increases faster than increases in the jurisdiction's property valuation.[11] Differential assessments reduce the valuation of property located in a jurisdiction. Therefore, efforts to reduce property tax burdens to *all* property owners are hampered by differential assessment policy; for a given budget, tax rates increase because PTA lowers the aggregate assessed value of property in the jurisdiction.

The extent of participation that can be expected in a preferential tax assessment program varies from state to state. In New Jersey, nearly 95 percent of the farmland is involved and the rate of participation across the state is not constant. In the suburban areas of Philadelphia and New York, Coughlin indicates that the rate of participation is well below the statewide average. It is important to note that the New Jersey program imposes extremely mild sanctions (two-year rollback of taxes with no interest penalty). In contrast, the California program is highly restrictive. Williamson contracts have terms that extend for ten years. Breach of the contract imposes a sanction of 12 percent of the market value of the land. Several studies have suggested that the California program, while containing substantial acreage, is not particularly successful in controlling the conversion of farmland in urbanizing areas of the state. As in New Jersey, California research shows that the rate of participation in urbanizing metropolitan areas is substantially below statewide levels of participation. Additionally, prime coastal areas exhibit low rates of participation and, due to the lack of planning and coordination, participant parcels are scattered throughout the state—obviously not highly effective in controlling sprawl.

Preferential tax assessment assumes that tax savings will be a strong inducement to prevent landowners from selling their property. Decisions to sell, while based to some extent on "tax squeeze" considerations, are more often the result of many factors—age of owner, the income-producing ability of the property, desire of children to assume the operation of the farm, desire to seek different work, as well as the negative effects brought on by urbanization that reduce profit. Clearly, many of the above factors have absolutely nothing to do with property tax considerations.

Coughlin's chapter assesses the effect of tax policies on rural land use. While data availability is limited, the author suggests that preferential taxation can generate two kinds of effects that influence farm ownership. First, by reducing property taxes, operating costs are reduced. Second, by increasing net farm revenues, the value of the farmland will increase through capitalization effects. This increase in value may reduce the gap between farm-derived market values and speculation-derived market values—thus allowing farm buyers entry to this market. The logic of this argument seems extremely weak, since the net effect is to increase overall land values. To the extent that higher farm revenues compensate for this effect, Coughlin may be correct. Relying on two separate studies of farmland and decline in Ohio, Coughlin suggests that only in cases where tax pressures are the key to forcing sales will PTA be effective. He also concludes that capitalization effects would not be sufficient to reduce the tendency or the ability of speculators to pick up farmland. Furthermore, in most states, speculators can enjoy the same tax advantages as farmers.

The costs of differential assessment, in addition to those outlined by Ladd, are the administrative costs of reviewing applications for participation, the policing of participants over contract periods and, most importantly, costs associated with foregone revenues—"tax expenditures." Tax shift costs can be substantial under circumstances where the extent of participation is significant relative to the tax base of the jurisdiction. They are, however, likely to be "modest" in urbanizing regions. Most shifts *are* modest but extremes have been recorded. In California, data suggest that tax losses varied from 4 percent to 22 percent of a county's tax. Several states have established provisions for transfers to local areas, to make up for the tax losses (California, New York).

In conclusion, Coughlin suggests that PTA responds to only one facet of the forces that predispose farmers to sell their land. As such, PTA is not highly effective in reducing the conversion of agricultural land. He suggests a variety of other planning devices that can control conversion: agricultural districting, public acquisition of farmland, and the purchase of development rights. Direct control over land use is necessary in areas that presently experience development pressure. PTA is not likely to be of much help in areas where conversion is eminent. If PTA is relied on to lessen the tax effects of urbanization, severe sanctions should be imposed on those who abrogate their contracts. However, as penalties increase, participation rates decline.

Coughlin's calculations illustrate the potential tax saving associated with PTA participation and the resulting tax shift that can be expected from the reassessment of agricultural land within the community. While far from a cost-benefit assessment of PTA, his illustrative calculations indicate that preferential programs may not be cost effective.

Operationalizing preferential taxation (Chapter 4)[12]

Up to this point we have focused on the ability of PTA to reduce the tax bill confronting owners of farmland. Banta addresses the implementation of preferential taxation programs. To delineate operational problems, he relies on four cases: New York, California, New Jersey and Oregon.

Several facets are essential to PTA implementation: eligibility criteria, an assessment method, and a monitoring system to ensure contract compliance. Specific issues that must be resolved are: (1) how use values are to be computed (from annual income statements, soils characteristics, or aggregate farm income figures); (2) whether detailed records of use and market value are to be kept; (3) whether tax base losses are to be compensated by subventions from other levels of government or whether tax rates within the jurisdiction are to be adjusted; and (4) what criteria will be used to determine eligibility.

While variation among states is common, several operational problems are generic. The choice of an appropriate capitalization rate of translating annual farm income into use values is a persistent problem. Controversies over the actual method of valuation are still prevalent in many states. Given the ad hoc methods often used to compute use value, variation in assessment levels is a serious problem plaguing tax officials. Banta suggests that uniform methods of use-value assessment be adopted in order to avoid problems of inconsistency within tax districts.

Most states using preferential taxation of agricultural lands view it as a reward for maintaining land in farm use. Other states more aggressive in preserving farmland are beginning to rely on other complementary techniques. These techniques are voluntary agreements, regulation prohibiting agricultural land conversion, and the purchase of development rights. Voluntary agreements have not yet gained widespread acceptance.

Where contracts are linked to zoning ordinances, differences between voluntary agreements and contracts appear. Such policy links allow for the easy monitoring and enforcement of preferential taxation eligibility requirements. The tying of zoning to the tax program often avoids the direct granting of compensation to injured parties.

Interest is growing for the use of development rights acquisition. Like zoning, development rights acquisition can be tied to preferential tax programs. Linkage problems develop when the residual value of the property, that is, after the development rights valuation is subtracted from the market value, does not correspond with the use value. Such problems seem to be associated with methods of accounting and valuation, and after the assessment procedures are corrected, the conceptual approach should still remain valid.

The final approach suggested by Banta is that of exclusive farm use zoning. This approach has merit because it avoids costs associated with development rights acquisition. Drawing on the Oregon experience, Banta suggests that such restrictive zoning approaches can be operationalized without undue legal challenges. By automatically tying preferential taxation to exclusive farm use zoning, the prospects for successful, legal challenges are minimized. Banta alleges that the expectations of property owners in Oregon are changing as a result of the EFU zoning and assessment programs.

The operational problems of preferential taxation of farmland and use are substantial. The fundamental problem seems to be the difficulty of determining use value in a way that is administratively feasible and promotes horizontal equity. Despite administrative problems Banta concludes that PTA, as an adjunct to other forms of land regulation or control, may prove to be useful.

If preferential taxation is so bad, why is it so widely used? (Chapter 5)[13]

The diffusion of public policy is complex. In the present context, preferential taxation of agricultural lands has spread to more than 40 states since 1956. Each adoption of PTA is characterized by different settings, actors and political groups. To unravel these complex factors Brigham uses an historical approach. He hypothesizes that preferential taxation is the result of de facto preferences for agriculture, extralegal tax preferences, and pressures brought on by the movement to 100 percent valuation of property. A corollary to his hypothesis is that the link between PTA and land-use policy is mythical, but wisely popularized by policy-makers.

Due to a variety of factors, pressures for 100 percent assessment developed. Movement to full valuation generated significant tax burdens on farmers. In response to such pressure, states began to enact preferential taxation schemes to ameliorate farm tax burdens. Most of these legislative changes occured well before the ascendency of open space or environmental rationales for preferential taxation of agricultural land. Brigham essentially argues that while the rationales (preservation of agricultural land and open space) served to garner legislative support, the shift to 100 percent assessment levels and rapid suburban land conversion "stand as the basis for interpretation of the preferential taxation phenomenon."

In most states that adopted PTA, state agricultural extension stations conducted research outlining the benefits associated with PTA. Proponents of tax schemes point out the relative desirability of PTA over direct intervention in the land markets (development rights acquisition, zoning, etc.). The subsidy approach appears to be preferred to regulation, and this ideology seems to be consonant with longstanding government relationships with agriculture. While the U.S. Department of Agriculture spent considerable efforts to evaluate the effects (i.e. the benefits) of PTA, little attention was given to the ability of tax schemes to alter farmer behavior—the principal line of attack taken by PTA critics.

PTA is perhaps better considered a tax subsidy scheme than a land-use policy scheme. It is only infrequently integrated with the land policy of a local community. Furthermore, as most critics point out, it does not alter the propensity of farmers to sell their land when confronted with the opportunity.

Conclusions

Preferential assessment of farmland is not an effective tool for controlling land use. Evidence suggests that it does not alter appreciably the conversion of agricultural land. Tax considerations are only a facet of the farmer's decision to sell out. Therefore, since PTA addresses only tax considerations, it is not capable of substantially altering the decision outcome. Administrative problems associated with PTA seem to be fairly serious and more intractable than those associated with market value assessment. From equity and efficiency perspectives, PTA does not perform well. It is a blunt policy instrument and it generates benefits that appear to be regressive. Land use planning in most states is not well integrated with tax preference schemes.

It appears that there are possibilities for linking PTA with other direct forms of government regulation of land. Banta's discussion of Oregon's program is suggestive. It may be unreasonable to assume, however, that such marriages can be arranged. Perhaps, as Brigham suggests, PTA is successful not because it achieves no popularly recognized land policy goals but because it generates substantial tax subsidies to agriculture.

Notes

1. John C. Keene et. al., *Untaxing Open Space*, Prepared for the Council on Environmental Quality (Washington, D.C.: G.P.O., 1976).

2. Barclay Hudson, "Domains of Evaluation," *Social Policy*, Vol. 6 (Sept/Oct 1975), pp. 79–89.

3. A. J. Harrison, *Economics and Land Use Planning* (St. Martin's Press, 1977), pp. 62–84.

4. For an attempt, see: Real Estate Research Corporation, *The Costs of Sprawl* for the Council on Environmental Quality, H.U.D., and E.P.A. (Washington, D.C.: G.P.O.), 1974.

5. For a thorough discussion of the optimal timing of land conversion see: Donald Shoup, "The Optimal Timing of Urban Land Development," Regional Science Association Papters, Vol. 25 (1970), pp. 33–44.

6. Helen F. Ladd, "The Tax Policy Considerations Underlying Preferential Tax Treatment of Open Space and Agricultural Lands," Chapter 2, this volume.

7. This definition follows from R. Musgrave and P. Musgrave, *Public Finance in Theory and Practice*, 2nd ed. (New York: McGraw-Hill, 1976). Chapter 10.

8. Ladd, op. cit., p. 19.

9. Generally, see: Robert Gloudemans, *Use-Value Assessment: Theory, Practice, Impact* (Chicago: International Association of Assessing Officials, 1974).

10. Robert Coughlin, "The Magnitude of the Savings and its Effect on the Land Market," Chapter 3, this volume.

11. For an excellent study of the impact of suburbanization on taxes, see Dick Netzer, "Financing Suburban Development," reprinted in *Municipal Needs, Services and Financing: Readings on Municipal Expenditures*, W. Patrick Beaton, ed. (Rutgers, N.J. Center for Urban Policy Research, 1974).

12. John S. Banta, "Operationalizing Differential Assessment for Agricultural Land," Chapter 4, this volume.

13. John Brigham, "The Politics of Tax Preferences for Farmland and Open Space," Chapter 5, this volume.

INDEX